APPLIED SOFT COMPUTING
Techniques and Applications

Research Notes on Computing and Communication Sciences

APPLIED SOFT COMPUTING
Techniques and Applications

Edited by
Samarjeet Borah, PhD
Ranjit Panigrahi, PhD

First edition published 2022

Apple Academic Press Inc.
1265 Goldenrod Circle, NE,
Palm Bay, FL 32905 USA
4164 Lakeshore Road, Burlington,
ON, L7L 1A4 Canada

CRC Press
6000 Broken Sound Parkway NW,
Suite 300, Boca Raton, FL 33487-2742 USA
2 Park Square, Milton Park,
Abingdon, Oxon, OX14 4RN UK

© 2022 by Apple Academic Press, Inc.

Apple Academic Press exclusively co-publishes with CRC Press, an imprint of Taylor & Francis Group, LLC

Reasonable efforts have been made to publish reliable data and information, but the authors, editors, and publisher cannot assume responsibility for the validity of all materials or the consequences of their use. The authors, editors, and publishers have attempted to trace the copyright holders of all material reproduced in this publication and apologize to copyright holders if permission to publish in this form has not been obtained. If any copyright material has not been acknowledged, please write and let us know so we may rectify in any future reprint.

Except as permitted under U.S. Copyright Law, no part of this book may be reprinted, reproduced, transmitted, or utilized in any form by any electronic, mechanical, or other means, now known or hereafter invented, including photocopying, microfilming, and recording, or in any information storage or retrieval system, without written permission from the publishers.

For permission to photocopy or use material electronically from this work, access www.copyright.com or contact the Copyright Clearance Center, Inc. (CCC), 222 Rosewood Drive, Danvers, MA 01923, 978-750-8400. For works that are not available on CCC please contact mpkbookspermissions@tandf.co.uk

Trademark notice: Product or corporate names may be trademarks or registered trademarks and are used only for identification and explanation without intent to infringe.

Library and Archives Canada Cataloguing in Publication

Title: Applied soft computing : techniques and applications / edited by Samarjeet Borah, PhD, Ranjit Panigrahi, PhD.
Other titles: Applied soft computing (2022)
Names: Borah, Samarjeet, editor. | Panigrahi, Ranjit, 1979- editor.
Description: First edition. | Series statement: Research notes on computing and communication sciences | Includes bibliographical references and index.
Identifiers: Canadiana (print) 2021026294X | Canadiana (ebook) 20210263024 | ISBN 9781774630297 (hardcover) | ISBN 9781774639238 (softcover) | ISBN 9781003186885 (ebook)
Subjects: LCSH: Soft computing.
Classification: LCC QA76.9.S63 A67 2022 | DDC 006.3—dc23

Library of Congress Cataloging-in-Publication Data

Names: Borah, Samarjeet, editor. | Panigrahi, Ranjit, 1979- editor.
Title: Applied soft computing : techniques and applications / edited by Samarjeet Borah, PhD, Ranjit Panigrahi, PhD.
Other titles: Applied soft computing (Apple Academic Press)
Description: First edition. | Palm Bay, FL : Apple Academic Press, 2022. | Series: Research notes on computing and communication sciences | Includes bibliographical references and index. | Summary: "Applied Soft Computing: Techniques and Applications explores a variety of modern techniques that deal with estimated models and give resolutions to complex real-life issues. Involving the concepts and practices of soft computing in conjunction with other frontier research domains, this book explores a variety of modern applications in soft computing, including bioinspired computing, reconfigurable computing, fuzzy logic, fusion-based learning, intelligent healthcare systems, bioinformatics, data mining, functional approximation, genetic and evolutionary algorithms, hybrid models, machine learning, meta heuristics, neuro fuzzy system, and optimization principles. The book acts as a reference book for AI developers, researchers, and academicians as it addresses the recent technological developments in the field of soft computing. Soft computing has played a crucial role not only with the theoretical paradigms but is also popular for its pivotal role for designing a large variety of expert systems and artificial intelligent-based applications. Beginning with the basics of soft computing, this book deeply covers applications of soft computing in areas such as approximate reasoning, artificial neural networks, Bayesian networks, big data analytics, bioinformatics, cloud computing, control systems, data mining, functional approximation, fuzzy logic, genetic and evolutionary algorithms, hybrid models, machine learning, meta heuristics, neuro fuzzy system, optimization, randomized searches, and swarm intelligence. This book is destined for a wide range of readers who wish to learn applications of soft computing approaches. It will be useful for academicians, researchers, students, and machine learning experts who use soft computing techniques and algorithms to develop cutting-edge artificial intelligence-based applications"-- Provided by publisher.
Identifiers: LCCN 2021032651 (print) | LCCN 2021032652 (ebook) | ISBN 9781774630297 (hardback) | ISBN 9781774639238 (paperback) | ISBN 9781003186885 (ebook)
Subjects: LCSH: Soft computing. | Computer science--Industrial applications.
Classification: LCC QA76.9.S63 A684 2022 (print) | LCC QA76.9.S63 (ebook) | DDC 006.3--dc23
LC record available at https://lccn.loc.gov/2021032651
LC ebook record available at https://lccn.loc.gov/2021032652

ISBN: 978-1-77463-029-7 (hbk)
ISBN: 978-1-77463-923-8 (pbk)
ISBN: 978-1-00318-688-5 (ebk)

RESEARCH NOTES ON COMPUTING AND COMMUNICATION SCIENCES

EDITOR-IN-CHIEF

Dr. Samarjeet Borah
Department of Computer Applications,
Sikkim Manipal Institute of Technology,
Sikkim Manipal University (SMU),
Majhitar, East Sikkim-737136, India
Email: samarjeet.b@smit.smu.edu.in
samarjeetborah@gmail.com

Brief Description of the Series

Computing can be defined as the practice in which computer technology is used to do a goal-oriented assignment. It covers design and development of hardware and software systems for various purposes. Computing devices are becoming an integral part of life now-a-days, including desktops, laptops, hand-held devices, smartphones, smart home appliances, etc. The evolution of the Internet of Things (IoT) has further enriched the same. The domain is ever growing and opening up many new endeavors, including cloud computing, social computing, ubiquitous computing, parallel computing, grid computing, etc.

In parallel with computing, another field has emerged that deals with the interconnection of devices. It is communication, and without which, the modern world cannot be thought of. It works with a basic purpose of transferring information from one place or person to another. This technology has a great influence in modern day society. It influences business and society by making the interchange of ideas and facts more efficient. Communication technologies include the Internet, multimedia, e-mail, telephone, and other sound-based and video-based communication means.

This new book series consists of both edited volumes as well as selected papers from various conferences. Volumes of the series will contain the latest research findings in the field of communication engineering, computer science and engineering, and informatics. Therefore, the books cater to the needs of researchers and readers of a broader spectrum.

Coverage & Approach

The series

- Covers a broad spectrum of research domains
- Presents on market-demanded product-based research works
- Discusses the latest developments in the field

The book series broadly considers contributions from the following fields:

- Artificial Intelligence and Expert Systems
- Big Data Analytics
- Broadband Convergence System and Integration Technologies
- Cellular and Mobile Communication
- Cloud Computing Technologies
- Computational Biology and Bioinformatics
- Computer and Information Security
- Computer Architecture
- Computer Graphics and Video Processing
- Control Systems
- Database Management Systems
- Data Mining
- Design Automation
- Digital Signal Processing
- GSM Communication
- High Performance Computing
- Human-Computer Interaction
- IoT and Blockchains
- Machine Learning
- Natural Language Processing
- Next Generation Communication Technologies
- Operating Systems & Networking
- Pervasive Computing and Cyber-Physical Systems
- Robotics and Automation
- Signal Processing

- Smart Internet of Everything
- SOC and System Platform Design Technologies
- Social Network Analysis
- Soft Computing

Types of Volumes

This series presents recent developments in the domains of computing and communications. It will include mostly the current works and research findings, going on in various research labs, universities and institutions and may lead to development of market demanded products. It reports substantive results on a wide range of computational approaches applied to a wide range of problems. The series provides volumes having works with empirical studies, theoretical analysis or comparison to psychological phenomena. The series includes the following types of volumes:

- Conference Proceedings
- Authored Volumes
- Edited Volumes

Volumes from the series must be suitable as reference books for researchers, academicians, students, and industry professionals.

To propose suggestions for this book series, please contact the book series editor-in-chief. Book manuscripts should be minimum 250–500 pages per volume (11 point Times Roman in MS-Word with 1.5 line spacing).

Books and chapters in the series are included in Google Scholar and selectively in Scopus and possibly other related abstracting/indexing services.

BOOKS IN THE RESEARCH NOTES ON COMPUTING AND COMMUNICATION SCIENCES SERIES

- **Applied Soft Computing: Techniques and Applications**
  Editors: Samarjeet Borah and Ranjit Panigrahi
- **Intelligent System Algorithms and Applications in Science and Technology**
  Editors: Sunil Pathak, Pramod Kumar Bhatt, Sanjay Kumar Singh, Ashutosh Tripathi, and Pankaj Kumar Pandey
- **Intelligent IoT Systems for Big Data Analysis: Concepts, Applications, Challenges, and Future Scope**
  Editors: Subhendu Kumar Pani, Pani Abhay Kumar, Samal Puneet Mishra, Ruchi Doshi, and Tzung-Pei Hong
- **Computing and Communications Engineering in Real-Time Application Development**
  Editors: B. K. Mishra, Samarjeet Borah, and Hemant Kasturiwale
- **Detection and Prediction of Serial Crime via Fuzzy Multi-Criteria Decision-Making Approach**
  Editors: Soumendra Goala and Palash Dutta

ABOUT THE EDITORS

Samarjeet Borah, PhD
Professor, Department of Computer Applications, SMIT, Sikkim Manipal University (SMU), Sikkim, India

Samarjeet Borah, PhD, is currently working as a Professor and Head in the Department of Computer Applications, SMIT, Sikkim Manipal University (SMU), Sikkim, India. Dr. Borah has carried out various funded projects from AICTE (GoI), DST-CSRI (GoI) etc. He has organized various workshops and conferences at national and international levels. Dr. Borah is involved with various book volumes and journals of repute from Springer, IEEE, Inderscience, and IGI Global as Editor or Guest Editor. He is the editor-in-chief of the book and proceedings series Research Notes on Computing and Communication Sciences, published by Apple Academic Press, USA. His areas of research are data mining, data science, and machine learning.

Ranjit Panigrahi, PhD
Assistant Professor, Department of Computer Applications at Sikkim Manipal University (SMU), Sikkim, India

Ranjit Panigrahi, PhD, is currently working as an Assistant Professor in the Department of Computer Applications at Sikkim Manipal University (SMU), Sikkim, India. His research interests are machine learning, pattern recognition, and wireless sensor networks. Dr. Panigrahi is actively involved in various national and international conferences of repute. He serves as a member of the technical review committee for various international journals of Inderscience and Springer Nature. He received his MTech in Computer Sciences and Engineering from Sikkim Manipal Institute of Technology and his PhD in Computer Applications from Sikkim Manipal University, India.

CONTENTS

Contributors .... *xv*
Abbreviations .... *xix*
Preface .... *xxiii*
Introduction .... *xxv*

1. **Swarm Intelligence and Bio-Inspired Computation** 1
   Rebika Rai

2. **User vs. Self-Tuning Optimization: A Case Study on Image Registration** 23
   Jose Santamaria Lopez and Maria L. Rivero-Cejudo

3. **Secure Communication Using a Novel 4-D Double Scroll Chaotic System** 41
   Pushali Trikha and Lone Seth Jahanzaib

4. **Detecting Hate Speech Through Machine Learning** 59
   F. H. A. Shibly, Uzzal Sharma, and H. M. M. Naleer

5. **Optimization of Logical Resources in Reconfigurable Computing** 69
   S. Jamuna

6. **A Sophisticated Similarity Measure for Picture Fuzzy Sets and Their Application** 89
   Palash Dutta

7. **Semi-Circular Fuzzy Variable and Its Properties** 105
   Palash Dutta

8. **Virtual Machine Selection Optimization Using Nature-Inspired Algorithms** 121
   R. B. Madhumala and Harshvardhan Tiwari

9. **Extractive Text Summarization Using Convolutional Neural Network** 135
   Mihir, Chandni Agarwal, Sweta Agarwal, and Udit Kr. Chakraborty

10. **Theory, Concepts, and Applications of Artificial Neural Network** 153

   P. Anirudh Hebbar, M. V. Manoj Kumar, and Archana Mathur

11. **Comparing Word Embeddings on Authorship Identification** 177

   Tarun Kumar Dugar, S. Gowtham, and Udit Kr. Chakraborty

12. **Fusion-Based Learning Approach for Predicting Diseases in an Earlier Stage** 195

   K. Krishna Prasad, P. S. Aithal, A. Jayanthiladevi, and Manivel Kandasamy

13. **A Fuzzy-Based Framework for an Agriculture Recommender System Using Membership Function** 207

   R. Narmadha, T. P. Latchoumi, A. Jayanthiladevi, T. L. Yookesh, and S. Prince Mary

14. **Implying Fuzzy Set for Computing Agricultural Vulnerability** 225

   A. Jayanthiladevi, L. Devi, R. Kannadasan, Ved P. Mishra, Piyush Mishra, and A. Mohamed Uvaze Ahamed

15. **Modeling an Intelligent System for Health Care Management** 237

   A. Jayanthiladevi, P. S. Aithal, K. Krishna Prasad, and Manivel Kandasamy

Index *249*

CONTRIBUTORS

Chandni Agarwal
Department of Computer Science and Engineering, Sikkim Manipal Institute of Technology, Sikkim Manipal University, Sikkim, India

Sweta Agarwal
Department of Computer Science and Engineering, Sikkim Manipal Institute of Technology, Sikkim Manipal University, Sikkim, India

A. Mohamed Uvaze Ahamed
Department of Information Technology, Qala University College, Kurdistan Region, Iraq

P. S. Aithal
Srinivas University, Mangalore, Karnataka, India

Udit Kr. Chakraborty
Department of Computer Science and Engineering, Sikkim Manipal Institute of Technology, Sikkim Manipal University, Sikkim, India, E-mails: udit.kc@gmail.com; udit.c@smit.smu.edu.in

L. Devi
PGCS Department, Muthayammal College of Arts and Science, Rasipuram, Tamil Nadu, India

Tarun Kumar Dugar
Department of Computer Science and Engineering, Sikkim Manipal Institute of Technology, Sikkim Manipal University, Sikkim, India

Palash Dutta
Department of Mathematics, Dibrugarh University, Dibrugarh – 786004, Assam, India,
E-mail: palash.dtt@gmail.com

S. Gowtham
Department of Computer Science and Engineering, Sikkim Manipal Institute of Technology, Sikkim Manipal University, Sikkim, India

P. Anirudh Hebbar
Department of Information Science and Engineering, Nitte Meenakshi Institute of Technology, Bangalore, Karnataka, India

Lone Seth Jahanzaib
Department of Mathematics, Jamia Millia Islamia, New Delhi, India,
E-mail: lone.jahanzaib555@gmail.com

S. Jamuna
Professor, Department of Electronics and Communication Engineering,
Dayananda Sagar College of Engineering, VTU, Bangalore, Karnataka, India,
E-mail: jamuna-ece@dayanandasagar.edu

A. Jayanthiladevi
Computer Science and Information Science, Srinivas University, Mangalore, Karnataka, India,
E-mail: drjayanthila@gmail.com

Manivel Kandasamy
Optum Tech-United Health Group, Bangalore, Karnataka, India

R. Kannadasan
School of Computer Science and Engineering (SCOPE), Vellore Institute of Technology (VIT) University, Vellore – 632014, Tamil Nadu, India

M. V. Manoj Kumar
Department of Information Science and Engineering, Nitte Meenakshi Institute of Technology, Bangalore, Karnataka, India, E-mail: manojmv24@gmail.com

T. P. Latchoumi
Department of CSE, VFSTR (Deemed to be University), Andhra Pradesh, India

Jose Santamaria Lopez
Department of Computer Science, University of Jaen, Spain, E-mail: jslopez@ujaen.es

R. B. Madhumala
Department of Computer Science Engineering, JAIN (Deemed to be University), Bangalore, Karnataka, India, E-mail: madhumala8887@gmail.com

S. Prince Mary
Department of CSE, Sathyabama Institute of Science and Technology, Chennai, Tamil Nadu India

Archana Mathur
Department of Information Science and Engineering, Nitte Meenakshi Institute of Technology, Bangalore, Karnataka, India

Mihir
Department of Computer Science and Engineering, Sikkim Manipal Institute of Technology, Sikkim Manipal University, Sikkim, India

Piyush Mishra
Department of Computer Engineering and Applications, GLA University, Mathura, Uttar Pradesh, India

Ved P. Mishra
Department of Computer Science and Engineering, Amity University, Dubai, UAE

H. M. M. Naleer
Senior Lecturer, South Eastern University of Sri Lanka, Sri Lanka

R. Narmadha
Department of ECE, Sathyabama Institute of Science and Technology, Chennai, Tamil Nadu, India

K. Krishna Prasad
Computer Science and Information Science, Srinivas University, Mangalore, Karnataka, India, E-mail: krishnaprasadkcci@srinivasuniversity.edu.in

Rebika Rai
Department of Computer Applications, Sikkim University, Sikkim, India E-mail: rrai@cus.ac.in

Maria L. Rivero-Cejudo
Department of Computer Science, University of Jaen, Spain

Uzzal Sharma
Assistant Professor, Assam Don Bosco University, Assam, India

F. H. A. Shibly
PhD Research Scholar, Assam Don Bosco University/South Eastern University of Sri Lanka, Sri Lanka, E-mail: shiblyfh@seu.ac.lk

Harshvardhan Tiwari
Professor, CIIRC, Jyothy Institute of Technology, Bangalore, Karnataka, India

Pushali Trikha
Department of Mathematics, Jamia Millia Islamia, New Delhi, India

T. L. Yookesh
Department of Mathematics, VFSTR (Deemed to be University), Andhra Pradesh, India

ABBREVIATIONS

ABC	artificial bee colony
ABCO	artificial bee colony optimization
ACO	ant colony optimization
AES	advance encryption standard
AFR	aggregated fuzzy ratings
AI	artificial intelligence
AIM	adaptive invasion-based model
AIS	artificial immune system
ALU	arithmetic and logic unit
AME	accumulated median error
ANC	artificial neural computing
ANNs	artificial neural networks
AS	ant system
ATSP	asymmetric traveling salesperson problem
BASK	binary amplitude shift keying
BBO	biography-based optimization
BCO	bee colony optimization
BFSK	binary frequency-shift keying
BLEU	bilingual evaluation understudy
BPSK	binary phase-shift keying
BPSO	binary PSO
CBOW	common bag of words
CLB	configurable logic block
CM	credibility measure
CNNs	convolutional neural networks
CSO	cat swarm optimization
CSP	candidate solution pool
CT	credibility theory
CV	computer vision
dDE	distributed DE
DE	differential evolution

DPSO	discrete PSO algorithm
DRC	design rules check
EAs	evolutionary algorithms
EC	evolutionary computation
EP	evolutionary programming
ERBD	evolutionary rigid-body docking
ESs	evolution strategies
EV	expected value
FFO	firefly optimization
FPAB	flower pollination by artificial bees
FPGAs	field-programmable gate arrays
FRS	fuzzy-based recommendation system
FS	fuzzy systems
FST	fuzzy set theory
FV	fuzzy variable
GAs	genetic algorithms
GloVe	global vectors
GSA	gravitational search algorithm
HS	harmony search
I.C.	initial conditions
IaaS	infrastructure as a service
ICAP	internal configuration access port
ICD	inverse credibility distribution
ILA	integrated logic analyzer
IR	image registration
IWC	importance weight of criteria
LCA	league championship algorithm
LE	Lyapunov exponent
LRVs	linguistic rating variables
LSTM	long short-term memory
LWVs	linguistic weighting variables
MA	memetic algorithm
MF	membership function
ML	machine learning
MLP	multilayer perceptron
MOPSO	multi-objective PSO

MPI	message passing interface
NM	necessity measure
OL	orthogonal learning
OLDE	orthogonal learning DE
OTP	one time programmable
PaaS	platform as a service
PAVA	priority aware VM allocation
PFS	picture fuzzy set
PM	possibility measure
PPSO	perturbed PSO
PR	partial reconfiguration
PReLU	parametric ReLU
PS	processor section
PSO	particle swarm optimization
PV-DBOW	distributed bag of words version of paragraph vector
PV-DM	distributed memory version of paragraph vector
QAM	quadrature amplitude modulation
QPSK	quadrature phase-shift keying
ReLU	rectified linear unit
RIRE	retrospective image registration evaluation
RNN	recurrent neural network
RoAs	ratings of alternatives
RP	reconfigurable partition
RUB	rational upper bound
RUBV	rational upper bound of the variance
SaaS	software as a service
SAMPL	signal analysis and machine perception laboratory
SBAF	Saha Bora activation function
SC	soft computing
SCFV	semi-circular fuzzy variable
SI	swarm intelligence
SMs	similarity measures
SOC	system-on-chips
SOM	self-organizing maps
SoTA	state-of-the-art
TPFS	triangular PFSs

TrPFS	trapezoidal PFS
TSP	traveling salesperson problem
V	variance
VIO	virtual input-output
VMP	virtual machine placement
VNS	variable neighborhood search
WNFDM	weighted normalized fuzzy decision matrix
WoC	weights of criteria

PREFACE

Soft computing deals with estimated models and gives resolutions to complex real-life issues. Soft computing is lenient of fuzziness, ambiguity, and approximations. In fact, it is based on the working methodology of the human mind. The term was conceived by Prof. Lotfi Zadeh, who also developed fuzzy sets that have many applications nowadays. In the field of evolutionary computing and other application domains such as data mining and fuzzy logic, soft computing techniques play an inimitable role, where it successfully levers state-of-the-art computationally intensive and complex problems that have usually appeared to be obstinate for traditional mathematical approaches. Due to the ability to counter innovative problem domains, soft computing remains pivotal among researchers in the field of information and communication technology. Soft computing addresses and is limited to theoretical aspects of computing; applied soft computing augmented the integration of soft computing tools and techniques into both day-to-day and advanced applications.

This book is a distinctive attempt to signify modern techniques designed to represent, augment, and encourage multidisciplinary research in the field of soft computing. Involving the concepts and practices of soft computing with other frontier research domains, this book cherishes varieties of modern applications in soft computing in varieties of the domain such as bioinspired computing, reconfigurable computing, fuzzy logic, fusion-based learning, intelligent healthcare systems, bioinformatics, data mining, functional approximation, genetic, and evolutionary algorithms, hybrid models, machine learning, metaheuristics, neuro-fuzzy system, and optimization principles. The book acts as a reference book for AI developers, researchers, and academicians, as it addresses the recent technological developments in the field of soft computing. Since the book reflects the application of soft computing in many fields, therefore all the chapters will be independent of each other; thus, it will allow the reader to go through the chapter directly to their interest area.

—**Editors**

INTRODUCTION

According to Matthew Haughn, soft computing (SC) is the estimated calculation intended to provide appropriate solutions to complex computation problems. The soft computing principle is popularly referred to as computational intelligence. Therefore, the main aim of soft computing is to solve problems, which are hardware intensive and time-consuming.

The book deals with soft computing principles and practices, wherein Chapter 1 deals with bio-inspired computation principles. In this chapter, the major ideas and principles of swarm intelligence (SI) have been highlighted, focusing on ant colony optimization (ACO), which is one of the widely used soft computing principles. Another nature-inspired algorithm, such as a PSO, has been detailed in other chapters. For instance, particle swarm optimization (PSO), a branch of soft computing, is the point of attraction in Chapter 8. Different nature-inspired algorithms and their application for optimal selection and placement of virtual machines in a cloud-based environment have been discussed in Chapter 8. Various parameters, including memory utilization, bandwidth, RAM speed, computation time and cost, etc., are considered to understand the capabilities of all the approaches. Each algorithm has its own limitations, and to find a very optimal solution will require developing a new solution based on the specific application requirements.

Apart from traditional nature-inspired algorithms, many more computational intelligence techniques are available, which the human being inspired, adopted, and applied in several applications. Further, a case study has also been explored in Chapter 2 using self-tuned evolutionary algorithms (EAs). The chapter covers a brief review of the SoTA on IR and several models using the self-tuning approach. Then, a computational study focused on the comparison of the performance of several of these IR methods dealing with 3D reconstruction problems of range images has been carried out. The statistical results reported in this work have shown the outstanding performance achieved by those IR methods using the self-tuning model of optimization.

Soft computing techniques are also explored as a novel 4-D double scroll chaotic system with stable and unstable equilibrium points in

Chapter 3. The dynamical properties, such as Lyapunov exponents (LE), existence, and uniqueness of solution, dissipative, and symmetric character, its equilibrium points analysis, is the main theme of the chapter. The application of the attained synchronization has been displayed in secure communication with the help of an illustration.

Social media trolling and abused is the main concern for social network engineers nowadays. Researchers have yet to find effective methods to detect and eradicate hate speech. The main challenge is that hate speech such as trolling and abusing can be in the form of text or sound files. Chapter 4 emphasizes that; the detection of hate speech is possible by utilizing various kinds of machine learning (ML) techniques. The ensemble of the classifiers of RNN and the model of CBOW are some of the evidence supporting the fact that detection of hate speech is possible utilizing all the several techniques of machine learning even though it is considered to be very challenging.

After addressing social and other issues, the soft computing approaches are implemented for efficiently utilizing logical resources. Therefore, Chapter 5 explores the partial reconfiguration (PR) technique facilitating an optimized number of resources during the implementation of multiple functional blocks with the time-multiplexed resources. The chapter also considered three case studies with respect to the utilization of logical resources. Moreover, soft computing is not limited to machine learning only. Instead, the umbrella of soft computing also covers many statistical techniques such as correlation and other similarity checking principles. In this context, Chapter 6 encompasses a sophisticated similarity measure for picture fuzzy sets (PFS), which according to the authors, are a novel technique. The approach of similarity is further applied in the decision-making problem of the supply of goods as well as in other problems. Soft computing also plays a critical role in understanding semi-circular fuzzy variables (FV) and their properties. In this regard, Chapter 7 investigated SCFV via credibility theory (CT) following a detailed explanation of PM, NM, CM, CD, and ICD. The variance and RUVB of SCFV have been examined in detail.

Neural networks as a soft computing approach have been explored extensively in Chapter 10. The content of the chapter touches upon various forms of activation functions of neural networks. The final part of Chapter 9 focuses on discussing the application of ANN in real-life scenarios, *viz.*, healthcare, self-driving vehicles, agriculture, and identifying astronomical objects. The neural network also proved to be a great aid for preparing a summary of text data. The word-embedding techniques discussed in Chapter 9 helps to represent input documents. Further, the derived architecture has been fed into a CNN model for classification. Word embedding techniques are also used for authorship identification in Chapter 11, where the focus of the work is to estimate the performance of the word-embedding techniques. Subsequently, deep and artificial neural networks (ANNs) are used to classify authors.

A comprehensive soft computing technique, popularly known as fusion-based learning, is also uniquely addressed in the healthcare arena for predicting diseases. Chapter 12 investigates potential competency for using the fusion concept and learning model to improve healthcare analytics. In the machine learning process, complex functionality has been done for enhancing analytical performance. With this learning process, it may facilitate risks to identify prediction in a more appropriate manner. Similarly, Chapter 15 encompasses an intelligent system for health care management. The objective of this work is to model an intelligent system for computing the assistance provided by hospitals during an emergency condition.

Other than the healthcare arena, soft computing lays its footprint in challenging areas like agriculture. Though there exist various techniques for agricultural recommender systems, the fuzzy-based framework proved to be a great aid using membership functions (MF). A fuzzy-based framework for agricultural recommender systems has been discussed in Chapter 13, where a recommender system is modeled specifically for recognizing and predicting consumption of diverse agriculture-based. The fuzzy-based model may predict items that are consumed by every customer; therefore, farmers may produce items based on requirements in hand. Fuzzy schemes are also explored within Chapter 14 for computing agricultural vulnerability. The joint framework of fuzzy sets and Mamdani membership function helps to understand vulnerability, dangerousness, drought resistibility, and exposure to drought.

CHAPTER 1

SWARM INTELLIGENCE AND BIO-INSPIRED COMPUTATION

REBIKA RAI

Department of Computer Applications, Sikkim University, Sikkim, India, E-mail: rrai@cus.ac.in

ABSTRACT

The biologically instigated world that embraces the social animals is known to all and the way behavior of these social animals are being simulated to design several algorithms is advancing and being employed in several research areas. One of the behaviors of the social animals/social agents that emphasizes on the direct or indirect synchronization among themselves and with the environment is popular and mostly used one. This has led to the development of the new discipline in field artificial intelligence termed as swarm intelligence (SI) that deals with natural and artificial systems using the algorithm that mimics the actions of social agents. The several algorithms from the disciple SI has been used in to solve several problems of different research vicinity and this chapter focuses on exploring one of the algorithms; i.e., ant colony optimization (ACO) that mimics the behavior of the real ants to solve travelling salesman problem (TSP) and for image processing.

1.1 INTRODUCTION

The social insect allegory for resolving predicaments has turn out to be a promising topic in recent years, highlighting on stochastic structuring practice, building the key probabilistically to optimize the result related to any kind of problems. The biologically inspired planet comprising of social agents basically emphasizes on direct and indirect communications wherein the cooperative activities of agent act together locally with the environment further causing the consistent global pattern to materialize. This muddled gathering of individual heading in arbitrary tracks that tends to cluster together is known as Swarms. Swarm intelligence (SI) [4, 5] and bio-inspired computing largely have gain enormous significance and drawn several attention in almost every area of science and engineering over the last decades. An apparent idea can be inferred by exploring some of the well-known scientific repository available, showing that the interest aroused by this branch has been swelling remarkably over the last few years. SI practice has been pertinent and utilized in resolving and optimizing [1] a variety of problems ranging from robotic navigation [9] and path planning to image processing [10].

SI is a promising discipline in the field of artificial intelligence (AI) in the literature are particle swarm optimization (PSO), ant colony optimization (ACO), flower pollination by artificial bees (FPAB), biogeography-based optimization (BBO), bee colony optimization (BCO), cuckoo search, firefly algorithm, bacterial foraging optimization and many more. Undoubtedly, the main sway behind the commencement of SI [4, 5] is the extremely renowned ACO algorithm; however other algorithms such as PSO, BBO, and FPAB are also widely used. Such noteworthy accomplishment of SI discipline has led to the introduction of several novel methods by several researchers on a different rousing sources varying from socio-political behavioral pattern to behavioral patterns of animals and even on physical processes of real-world which showcases the potential and flexibility of the discipline as a problem solver and optimizer in various fields. In this chapter, an overview of one of the most widely used bio-inspired algorithms, i.e., ACO has been explored. Furthermore, this chapter will highlight the application of ACO to solve traveling salesman problem (TSP) and in the field of Image Processing to detect the edges of the imagery.

1.2 ARTIFICIAL INTELLIGENCE (AI)

Artificial intelligence (AI) is becoming an elementary component of business intensification across a broad range of industries. AI [2] is the branch of computer sciences that gives emphasis to the design and development of intelligent machines [13] that are capable of thinking, working, and reacting like humans. The major sub-fields or disciplines of AI include machine learning (ML), neural networks, evolutionary computation (EC)/SI, robotics, expert systems, speech processing, natural language processing, and planning. It is a science and a set of computational techniques that are instigated by the manner in which human beings use their nervous system and their body to feel, learn, reason, and act. Whenever we think of humans, the major characteristics that make human intelligent are: reasoning, learning, decision making, planning, self-correction, problem-solving, knowledge representation, perception, motion, manipulation, creativity, and not to forget emotions.

AI intends to come up with a machine that is intelligent at par with the natural intelligence displayed by humans, so AI is a debatable topic. AI has become a topic of controversy bigger than ever before. Many people are worried about robots taking over the world. This very idea scares the world as people have started believing that bots are being created to replace people in the near future, and there is though, no denying the fact. However, AI [13] has become an inherent element and has an enormous impact on our lifestyle with the crucial uses of mobile phones, the bots, texts, and speech translators, driver assistance, voice assistance and systems that assist in suggesting merchandises and services.

The applications of AI are endless and can and have been applied to many different sectors and industries. AI is the reason: (a) Tesla uses autopilot (advance driving assistance system) that enables lane centering, adaptive cruise control, self-parking, the ability to automatically change lanes, navigate autonomously on limited-access freeways, and the ability to summon the car to and from a garage or parking spot; (b) Netflix and YouTube can read viewers minds by analyzing the viewing habits of the subscribers and create clusters known as taste communities; (c) Healthcare industry uses AI for dosing drugs and different treatment in patients, and for surgical procedures in the operating room; (d) Financial industry uses AI to detect and flag activity in banking and finance such as unusual debit card usage and large account deposits thereby providing all help a bank's fraud department; (e) Gmail uses AI in addition to the rule-based filter for years. Rule-based blocks most of the spam, however, ML looks for new patterns and suggest an e-mail is not to be trusted and categorizes it as spam; (f) Facebook uses AI to detect content falling under the categories such as nudity, fake account, suicide prevention, graphic violence, terrorism, hate speech; (g) Several voice assistance such as Google Assistant, Siri, Alexa, Corona are available in the market and they are not just to reply queries but to perform task like placing an online order, searching a particular song, typing messages and the list goes on.

Today it will not at all be surprising if human so-called "the most intelligent species" would fail to differentiate a robot and a human. There are in fact many emotional robots available in the market today that has the capability and caliber to emotionally portray them the way human does. On the other hand, several challenges [13] are associated with AI, to name the few which has been tabulated in Table 1.1.

TABLE 1.1 Challenges Associated with Artificial Intelligence

Decisive Factor	**Description**
Case-specific learning	Human intelligence permits them to use their experience from one domain to another, from one context to another, from one field to another; from one scenario to another and there lies the problem in AI since it continues to face difficulties carrying out its experience from one set of circumstances to another making it specialized strictly for just specific task
Creating and building trust	AI is all about science, technology, and algorithms which most people are unaware of, which makes it difficult for them to trust it. The major problem is that it is like a black box for many people making them doubtful about how decisions are taken and whether all its decisions are perfect or not. Also, there is lack of understanding of AI among non-technical employees leading to fewer people supporting and accepting it
Data privacy and security	The majority of the applications of AI are based on an enormous amount of for the purpose of learning and making intellectual decisions. Machine learning (ML) systems depend on the data, which is often sensitive and personal in nature. ML systems can become prone to data breach and identity theft and the main reason behind could be the systematic learning.
Algorithm bias	A huge dilemma with AI systems is that their level of goodness or badness totally depends on the program/algorithm and the data they are trained on. Thereby if a bias that lies in the algorithms which take critical decisions goes unrecognized, leading towards disreputable and unfair consequences.

TABLE 1.1 *(Continued)*

Decisive Factor	Description
Data scarcity	Though organizations have access to numerous data but the availability of datasets that are applicable for AI applications to learn are rarely available. That is the reason behind training the AI machines using supervised learning mechanism making use of labeled data that has limits and scarcity.
Field specialist scarcity	Both technical knowledge, as well as business understanding, plays a crucial role in developing a thriving AI solution; however, there is a scarcity of such specialist. So the number of good data scientists and AI experts who are familiar and well versed with the knowledge of gracefully applying technology to a given business problem is limited in numbers.
Integration challenges	The very process of integrating AI into the existing systems is challenging that requires considering several components such as data storage, labeling, infrastructure needs, feeding data to the system, training, and testing the models developed, continuous monitoring of the developed models, feasibility of the resources, data sampling, quick processing and not to forget accurate outcomes.

1.3 INTRODUCTION TO SWARM INTELLIGENCE (SI)

Swarm is basically a large or dense group of disorganized collection or population of individual moving somewhere in random directions that tends to cluster together. Gerardo and Jing in 1989 coined the term SI that tends to mimic the behavior of the social animals or agents available in our nature such as foraging behavior of ants, flocking behavior of birds, fish schooling, bacteria molding, animals herding and many more [16]. SI basically emphasizes on direct and indirect interactions whereby the cooperative behavior of agent interacting locally with the environment causes the coherent global pattern to emerge.

SI [4] system possesses various properties, i.e., it is composed of many individuals; the individuals are relatively homogeneous (i.e., they are either all identical or they belong to a few typologies); the interactions among the individuals are based on uncomplicated behavioral laws which make use of only local information that the individuals exchange directly or via the environment (stigmergy); on the whole, behavior of the system is the consequence of the interactions of individuals with each other and with their environment, that is, the group self-organizes. However, the main illustrative and vital property of a SI system is its capability to perform in a coordinated way even in the absence of a coordinator or of an external regulator. For instance, the beautiful yet complex dance performed by the flock of birds in the sky. There are many features to be noted and analyzed that can be used and implemented in real-world applications. If birds are added to the flock, they continue to fly and scale in huge dimensions. However, if a bird falls on the ground, the flock crashes on the ground. Together they can do more than what they can when being alone. They can together avoid predators and the most excellent element is there is no leader in the group that directs the group concerning what and what not to do yet preserve such beautiful balance. It is the end result of every bird keeping an eye on their local neighbor following the uncomplicated set of rule called self-organization. There are examples of self-organization everywhere in nature, may it be ant creating trails on the picnic table or bees as they make decision on selecting their next site. Therefore, SI can be deliberated as of something like the one shown in Figures 1.1 and 1.2 that basically is trying to highlight the fact that, since the brain is a system of neuron which are intensely connected that intelligence forms, Swarm can be very well thought of system of brains so profoundly bonded that superintelligence forms, making it further additionally smarter and intelligent than an individual matter.

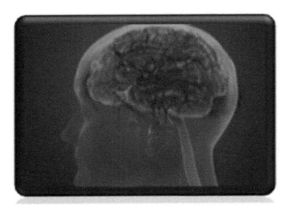

FIGURE 1.1 Individual (a brain).

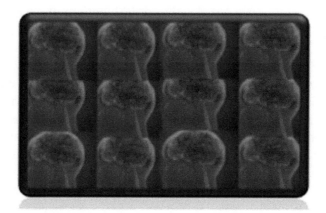

FIGURE 1.2 Swarm (system of brains).

1.4 SWARM INTELLIGENCE (SI) MODELS

The various SI models are available in the literature; however, we will primarily focus on one of the most popular SI modes [6] namely: ant colony optimization (ACO), PSO, FPAB, biogeography-based optimization (BBO), artificial bee colony optimization (ABCO), bacterial foraging optimization, cat swarm optimization (CSO), glowworm swarm optimization [5], and many more. ACO is a meta-heuristic approach for solving hard combinatorial problem inspired by the foraging behavior of ants proposed by Dorigo in the year 1992 [1] his doctoral dissertation. The inspiring sources of ACO are the pheromone trail placing [11] and following activities of real ants that utilizes pheromone as the contact medium. In correspondence to biological example, ACO algorithm deals with indirect interactions among the simple agents known as artificial ants mediated by artificial pheromone. Pheromone path acts as a distributed numerical data which ant uses to construct the probabilistic solution of problem. It is used during the algorithm implementation to replicate their exploration experience.

PSO is one of the popular population-based stochastic optimization technique inspired by societal behavior of bird flocking or fish schooling proposed by Eberhart, Kennedy, and Shi in 1995 [17]. PSO emulates the behavior of animal society that do not have leaders in their group and wherein the members tend to move in random pattern by following any one of its associate that has the closest position with the food resource. The velocity related with every particle directs the flying of particle. In PSO, each single solution in the search space is bird itself and is known as a "particle" that has an associated fitness values evaluated by the fitness function to be optimized. The particle follows the concept of exploration and exploitation. Exploration deals with the ability to move to different location in the search space in order to have the knowledge about the global best solution whereas exploitation is the ability to refrain the search within the confining area to gather the information about its local best solution.

FPAB is a stochastic optimization technique, proposed by Kazemian in 2006, inspired by the behavior of bees [18]. The major characteristics of bees taken into consideration are the capability of the bee to carry pollen and pollinate flowers in the search space. In analogous to the biological example, FPAB algorithm treats each bee as a simple artificial agent with a one-slot memory basically for the purpose of storing the development of pollen carried by each agent in its source. The bees select or choose the pollen of flower with lowest growth and pollinate the pollen where it survives better. After pollination, a variety of flowers with different growth rate or the same type of flowers with different growth rate is available in the search space. The flower with best growth of one species is allowed to survive while weeding out the ones with lower growth rate using a mechanism known as Natural Selection. Implementing the effortless activities with memory, the artificial bees can perform complicated tasks such as clustering.

1.5 ANT COLONY OPTIMIZATION (ACO)

The first thriving SI model is ACO, which was initiated by Dorigo [3], and has been innovatively used to solve discrete optimization problems. It is based on a study on a group of 'almost blind' ants mutually finding out the shortest route between their food and their nest with no visual information.

1.5.1 ANTS IN NATURE

Ants, as already mentioned earlier, communicate with each other using chemical essence known as pheromones, which is basically volatile in nature. Karlson and Lüscher in 1959 introduced for the very first time, the term "pheromone," based on the Greek word herein (means to transport) and hormone (means to stimulate) [19]. An ant perceives the intensity of the chemical on the ground with the help of their long, mobile antennae. There are various types of pheromones used in nature by different social agents, one being the food trail pheromone and another being alarm pheromone. Food trail pheromone basically is the chemical deposited by the ant on the surface as they move their way to search for the food source. This route will gradually attract other ants to follow [7, 9–11], thus marking the trail to pick up the shortest route from food source to their nest. On the other hand, alarm pheromone is produced by the crushed ants as a signal to alert nearby ants to either fight or escape predators and to protect their colony. The mechanism for real ant movements is illustrated in Figures 1.3(a)–(d). Ants communicate with each other using pheromones.

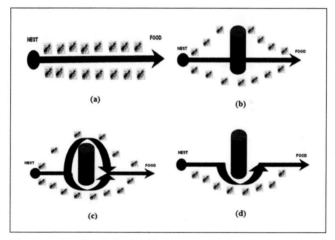

FIGURE 1.3 Ants foraging behavior [7, 10, 11]: (a) movement of ants from source (nest) to destination (food); (b) an obstacle placed on the way between nest and food; (c) ants randomly choosing the path; (d) shortest path chosen by maximum ants based on pheromone deposits.

A forager that finds food marks a trail on the way back to the colony so as to enable the other ants to follow, as shown in Figure 1.3(a). The ants then emphasize the trail when they head back with food to the colony. When the food source is exhausted, no new trails are marked by returning ants and the scent slowly dissipates [7, 10, 11]. This behavior helps ants to deal with changes in their environment. For instance, when an established path to a food source is blocked by an obstacle, as shown in Figure 1.3(b), the foragers leave the path to explore new routes as depicted in Figure 1.3(c). If an ant has accomplished its goal, it leaves a new trail marking the shortest route on its return. Successful trails are followed by more ants, as highlighted in Figure 1.3(d), reinforcing better routes and gradually finding the best path.

1.5.2 REAL ANTS VS. ARTIFICIAL ANTS

Comprehending a nature-inspired phenomenon and further scheming algorithm based on the comprehended phenomenon are correlated, yet diverse tasks. Understanding a natural phenomenon is controlled by examinations and testing, however designing an algorithm based on the understanding is totally based on one's thoughts and existing technology. The same has been illustrated in Figure 1.4.

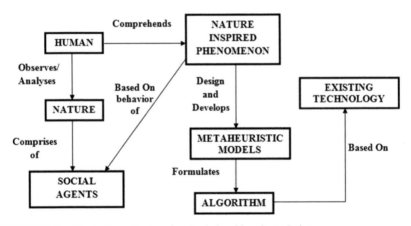

FIGURE 1.4 Illustration of nature-inspired algorithm formulation.

Although the ideology of the ACO algorithm is inspired by the foraging behavior of real ant, nonetheless, a number of characteristics of artificial ants do not have to be identically the same as real ants. Like real ants, the artificial ants are considered to have the following characteristics: artificial ants have memory; they are not completely blind, and they live in a discrete-time environment. However, they have some adopted characteristics from the real ants as: probabilistically prefer path with a larger amount of pheromone; shorter path is true path and larger is the rate of growth in the pheromone concentration; communicate to each other by means of the amount of pheromone that is laid on each path. Based on various decisive factors, Table 1.2 highlights few differences between real and artificial ants.

TABLE 1.2 Real Ants vs. Artificial Ants based on Various Decisive Factors

Decisive Factor	Real Ants	Artificial Ants
Nature of environment	Survives in Natural environment	Lives in discrete time, artificial/virtual environment,
Environmental constraints	Presence of predators, ants of other colonies	In the artificial environment, no such constraints exist
Memory slots	Real ants has no memory	Artificial ants has memory slots (limited) to trace the path
Visibility	Almost blind, perceives the intensity of pheromone using their long, mobile antenna	Aren't completely blind, traverses, and keep tracks of the path length in the attached memory module
Pheromone deposition	Both ways (forward and return path) pheromone deposition takes place	Only one way (return path) pheromone deposition takes place
Pheromone evaporation	Pheromone evaporates gradually making it less considerable for the convergence	Pheromone evaporates exponentially making it more considerable for the convergence
Pheromone updating amount	The more and more ants follows the path, pheromone concentration goes stronger	Once path is constructed, while returning the pheromone trail of that path is updated, the amount to be updated is inversely proportional to the path length stored in memory
Tracing the returning path	The quantity of pheromone deposited on their forward path is used	Uses the stored path available in the memory to trace the return path

It is significant to highlight that the presence of the memory slots is the key difference between real and artificial ants; the use of memory facilitates artificial ants to execute several constructive behaviors to proficiently construct results for intricate optimization predicaments. Artificial ants use this additional factor such as memory slots availability to evaluate solutions generated to formulate the quantity of pheromone to be deposited on its path. This being the main reason as to why the pheromone is only deposited on the return path after a full path or so-called candidate solution is build and assessed in context to path length as stored in memory.

1.6 ACO METAHEURISTIC

The inspiring resources of ACO are the pheromone trail laying and following behavior of real ants that uses pheromone as the communication medium that helps ants find the shortest route between their nest and a food source. ACO has resolved numerous optimization problems such as DNA sequencing, assembly line balancing, traveling salesman problem (TSP), scheduling, and many more. ACO is a relatively new population-based approach to problem solving that takes inspiration from the social behavior of the ants. Different ants take up different paths to reach the food source and deposit the substance on the ground known as pheromone-based on the fact that higher concentration of pheromone is deposited on shorter paths and minor concentration on longer paths. Here, the collective behavior of ants provides an intelligent solution for finding the shortest path from the nest to the food source. If a single ant finds a shorter path and leaves soaring concentration of pheromone on the way to the food source, then all the other ants' gets attracted towards it and hence following the shorter path. The crucial component required for the devising any ACO algorithm is the determination of the fitness function based on the decision of which the elements of a construction graph in regard to any problem will be provided with a high amount of pheromone trail, and to determine how ants will exploit these promising components when constructing new solutions. The important components in the ACO algorithm is highlighted in Table 1.3. The ACO algorithm designed based on ants' movement phenomenon is stated in Algorithm 1.1.

TABLE 1.3 Ant Colony Optimization Algorithm

Decisive Factor in ACO	Ant Colony Optimization (ACO)
Field	Artificial intelligence (swarm intelligence)
Type of algorithm	Nature-inspired metaheuristic
Nature of problem	Combinatorial optimization, continuous problem

TABLE 1.3 *(Continued)*

Decisive Factor in ACO	Ant Colony Optimization (ACO)
Objective of algorithm	Solving computational problems by searching for an optimal path through graphs (construction graph)
Computational agent/social agent involved	ACO is modeled on the actions of ants
Communication medium	Direct: A volatile chemical essence known as pheromones
	Indirect: Stigmergy (through the environment, between agents)
Communication mechanism	ACO use indirect communicate mechanism among ants and with the environment locally for the emergence of global pattern
Problem representation	Construction graph (weighted graph)
Applicability	For problem wherein source and destination are predefined and specific
Applications	Protein folding, credit risk assessment, vehicle routing, transportation systems, assembly line balancing, scheduling, traveling salesman problem (TSP), and image processing

Algorithm Name: Ant Colony Optimization (ACO) Algorithm [8, 11]
Algorithm Description: This algorithm highlights the general steps used in ACO.

Variable Description:
K: Number of ants; $\tau^{(0)}$: Initial pheromone matrix; k: ant index; n: construction step index; $p^{(n)}$: probabilistic transition matrix; L: construction step.

Input and Output:
Input: $\tau^{(0)}$; position of K ants; L
Output: $\tau^{(n)}$: final pheromone matrix.

Step 0: Start.
Step 1: Initialize the positions of K ants, and the Pheromone matrix $\tau^{(0)}$.
Step 2: [Loop] for the construction-step index n = 1: L.
Step 2.1: [Loop] for the ant index k = 1: K.
Step 2.1.1: Construct pheromone matrix and heuristic matrix.
[End Loop]
[End Loop]
Step 3: Consecutively move the Kth ant for L steps, according to a probabilistic transition matrix, $p^{(n)}$.
Step 4: Update the pheromone matrix $\tau^{(n)}$.
Step 5: Make the solution decision according to the final pheromone matrix $\tau^{(n)}$.
Step 6: End.

ALGORITHM 1.1 General steps of ACO.

1.7 ACO APPLICATIONS

ACO algorithms have been used in optimizing various combinatorial problems, ranging from protein folding to image processing to credit risk assessment to Vehicle routing to transportation systems to assembly line balancing to scheduling [12]. The algorithm has been utilized to generate near-optimal solution to problems such as the TSP, edge detection, feature extractions (image processing) since the ACO algorithm can run constantly and become accustomed to alterations in real-time. The foremost ACO algorithm termed as ant system (AS), intended to resolve TSP with an objective to plan a route through a number of cities such that each city is only visited once and with the salesman returning to his city of origin. To highlight the application of ACO, two scenarios, i.e., ACO in the field of Robotic navigation and path planning to solve the typical case of TSP; and ACO for detection of edges of imagery in the field of Image processing, has been highlighted briefly in the subsequent sections.

1.7.1 ACO TO SOLVE TRAVELLING SALESMAN PROBLEM (TSP)

SI approach emphasizes on direct or indirect interactions among simple agents and offers an alternative way of designing intelligent systems. To find the shortest path between the food (destination) and nest (source), without any visible, central, active coordination mechanism, ACO which is one of the popular approaches that searches for an optimal solution from among the given set of solutions can be applied to highlight how the group of agents or robots can perform navigation and plan the path accordingly. Feedback by the agent during traversal on the path results in more concentration on the path, thereby influencing the behavior of the other agents. Numerous obstacles are likely to be come across in the course of this traversal. The objective of the agent is to find an optimized solution to bring itself closer to the goal considering the cost, time, and path availability. A typical case of traveling salesman problem (TSP) [7, 9] can be considered to achieve it wherein an agent plans a route through a series of nodes (locations) and each location is just considered once with the agent returning back to city of origin.

The TSP [15] is a real-life business setback typically illustrated as scheduling a path through a number of cities such that each city is only

Swarm Intelligence and Bio-Inspired Computation

visited once and with the salesman arriving to his city of origin, i.e., the initial city from where he began his journey. The problem gets more complex as and when the number of cities grows up and a visual assessment no longer provides an apparent resolution, which entails demanding quantity of time, traveling cost as well as labor. A number of advances have been made with the introduction and implementation of new algorithms to resolve it, which on the other hand, is still costly for larger areas (as the number of cities that is to be traveled increases). In this regard, the necessity of reducing the overall cost-effectiveness is of paramount importance.

Initially, each of the 'K' ants is placed in randomly chosen cities and at each city a state transition rule is applied iteratively. Ants construct a tour solving the TSP using ACO as depicted in Figure 1.5.

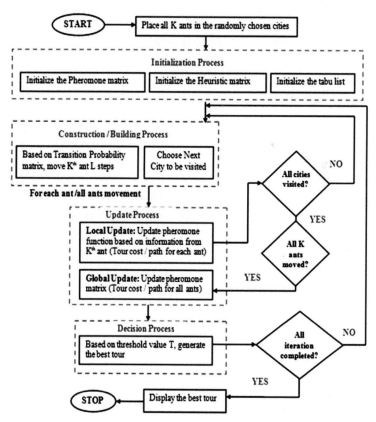

FIGURE 1.5 Flowchart for solving TSP using ant colony optimization.

The description of ants constructing the best tour is provided as follows: Assuming each ant 'k' has memory slot associated with a limited amount of memory termed as "tabu list" [15], initially consisting of current partial tour information and that gets updated at a regular interval with the movement of ants and the visited cities covered; At a particular city 'x,' ant probabilistically chooses still unvisited city 'y' (city which are close and connected by path with higher concentration of pheromone) based on the pheromone matrix and the locally available heuristic information from the heuristic matrix is taken into account to establish the path from city 'x' to 'y'; The "tabu list" is updated and is used to determine at each of the construction/building step the cities to be visited next to further build a realistic solution enhancing the ants to map out the tour once all the cities has been visited. Once each ant moves, the local update is performed by updating the pheromone matrix, and once all the cities are visited, and all ants have moved, the global update is performed wherein the Pheromone matrix is updated, generating the tour cost/path. This is characteristically performed by initially lessening the pheromone content/pheromone value by a constant factor and then later, as and when ants move and visit the path or arcs between cities, they tend to deposit the pheromone on it.

1.7.2 ACO FOR DETECTION OF EDGES OF AN IMAGERY

The decisive problem in image processing is to reveal constructive information, and it is scaling in recent years, fascinating investigators from various different fields to recommend techniques capable of effectively processing of an input image generating the desired output without compromising on the quality of the output. Image processing considers the image as a two-dimensional array of intensities of each pixel as an input and applies several mathematical operations to it. The various operations performed in image processing [14] include: image acquisition; image enhancement; image restoration; image compression; image segmentation; representation; and description, object recognition, and image classification. Edge is an important feature of an image which carries key information about the objects present in the image. One of the fundamental problems in image processing is the identification of sudden or abrupt changes in the brightness of an image, typically known as edge detection.

Edge detection [8], being one of the significant methods in Image Processing, basically provides the boundaries of object, thereby significantly reducing the amount of data to be processed. It weeds out the insignificant information while preserving the crucial structural properties of an image. Edge detection focuses on capturing the important events and change in properties due to discontinuities in surface orientation, material properties, or illumination discontinuities *viz;* surface normal discontinuity, depth discontinuity, surface color discontinuity and illumination discontinuity. The success of edge detector totally depends on the grit of threshold value. Various edge detection methods available in the literature [10] include Roberts, Sobel, Prewitt edge detectors, LoG edge detector, Marr-Hildreth edge detector, Canny edge detector, Basic declivity edge detector, Local threshold and Boolean function-based edge detector, etc. Several progresses have been made with the introduction and implementation of new algorithms to detect the edges of an imagery, and ACO [12], which is one of the swarm computing techniques too is paving its way towards the same.

The input image is considered as a 2-D graph where nodes are pixels of the image and ants are placed randomly at each pixel. Ants navigate from one node to another in the graph, thereby marking the trail by means of pheromone [11], and that represents an edge that captures the local transition in the intensity value as ant navigates. Each of the ants is assigned an arbitrary location in the image of size $M_1 \times M_2$ considering a total of K ants. The initial value of all elements is set to a small non-zero value in the pheromone matrix. The minimum pheromone intensity is set with the initial value so that the probability of choosing the pixel is never 0. The heuristic information matrix is constructed based on value captured at each edge of the graph, and as the movement takes place, the matrix gets updated. During the procedure using ACO, the pheromone matrix is updated twice. Firstly, local update after each Kth ant moves within the construction step and secondly, a global update [11] when all K ants move. Finally, an ant-based decision process is performed wherein a pheromone matrix is used to classify each pixel either as an edge or a non-edge by applying threshold T on the final pheromone matrix.

1.8 SWARM INTELLIGENCE (SI) ADVANTAGES AND LIMITATIONS

SI [4] provides a heuristic to solve difficult optimization problems with the basic philosophy of observing the behavior of animals in nature and then mimic those animals' behavior using algorithms. Because of its applicability in several areas such as medicine, military, scheduling, routing, telecommunication network, process optimization, etc., SI is now being adjudged as one of the most potential AI techniques with sturdy mounting scientific consideration with the increase in the number of successful SI research outputs. However, SI has some limitations that need to be looked after as a future perspective. The various advantages and limitations of SI have been tabulated in Tables 1.4 and 1.5, respectively.

TABLE 1.4 Advantages of Swarm Intelligence

Decisive Factors	Description
Scalability	SI systems are extremely scalable, with groups ranging from satisfactorily few up to millions of individuals. The control mechanisms used in SI systems are not too reliant on swarm size; the same control architecture can be applied to a couple of agents or thousands of agents.
Adaptability	SI systems retort healthily to speedily varying environments, beautifully, and swiftly exercising their self-organization potential at run time with considerable flexibility.
Fault tolerance	SI systems cooperatively work without central control, and there is no single individual crucial for the swarm to continue to function. As we are aware, a single point of failure is a part of any system that puts the entire system at risk of a complete failure, the fault tolerance potential of SI systems is outstandingly high, since these systems have no single point of failure.
Individual simplicity	SI systems consist of a several simple individuals with fairly limited capabilities on their own, yet the simple behavioral rules at the individual level are practically sufficient to cooperatively emerge sophisticated group behavior.
Speed	Alterations in the network can be propagated quickly.
Modularity	Agents act independently of other network layers.
Autonomy	Little or no human supervision is required.
Parallelism	Agents operations are inherently parallel.
Flexibility	The agents can be added or removed without hampering or influencing the structure.

TABLE 1.5 Limitations of Swarm Intelligence

Decisive Factors	Description
Time-restricted applications	The pathways to generate the solution for a problem using SI are basically emergent and not pre-defined, so SI is not suitable for time critical applications.
Tuning of the parameters	Parameters of SI systems are problem dependent so heuristically assumed, either empirically pre-selected (trial and error basis) or adaptively adjusted at run-time. So, tuning the parameters is one of the limitations.
Premature convergence	Due to the lack of central coordination, SI system are highly redundant and might endure from premature convergence and get trapped in local minimum.
Uncontrollable	Difficult to exercise control over swarm.
Non-immediate	SI systems are extremely scalable with groups ranging from satisfactorily few up to millions of individuals so; the more complex is the swarm, the longer it takes to shift states.

1.9 CONCLUSION

In this chapter, the major ideas and principles of SI has been highlighted, focusing on one of the most widely used bio-inspired algorithms, i.e., ACO. Nature is filled with inspiring sources and humans have from last many years started mimicking and taking advantage of the behavior of the social agents available in the nature to resolve real-life problems. Social insect's biology has found a solution to hard computational problems. SI basically relies on two important principles, namely self-organization, which is based on activity amplification by positive feedback; activity balancing by negative feedback and Stigmergy, which is based on stimulation by work; behavioral response to the environmental state are widely employed to solve various problem of real-life scenario. Various SI-based computational models are available and widely used with several successful applications on different areas yet having very high potential to solve many more that are yet to be explored in the near future.

1.10 KEY TERMINOLOGY AND DEFINITIONS

1. **Artificial Intelligence (AI):** It is the simulation of human intelligence processes by machines, especially computer systems.

These processes include learning (the acquisition of information and rules for using the information), reasoning (using rules to reach approximate or definite conclusions), and self-correction.

2. **Swarm Intelligence (SI):** It is the collective behavior of decentralized, self-organized systems, natural or artificial. The concept is employed in work on AI. The expression was introduced by Gerardo and Jing in 1989, in the context of cellular robotic systems [16].

3. **Metaheuristic:** It is a higher-level procedure or heuristic designed to find, generate, or select a heuristic (partial search algorithm) that may provide a sufficiently good solution to an optimization problem, especially with incomplete or imperfect information or limited computation capacity. Metaheuristic sample a set of solutions which is too large to be completely sampled. Metaheuristic may make few assumptions about the optimization problem being solved, and so they may be usable for a variety of problem.

4. **Ant Colony Optimization (ACO):** It is a population-based metaheuristic that can be used to find approximate solutions to difficult optimization problems. In ACO, a set of software agents called artificial ants search for good solutions to a given optimization problem.

5. **Image Processing:** It is a method to perform some operations on an image, in order to get an enhanced image or to extract some useful information from it.

6. **Robotic Navigation and Path Planning:** It means the robot's ability to determine its own position in its frame of reference and then to plan a path towards some goal location. Also, it aims to search for a safe path in a cluttered environment for a mobile robot from source to destination, avoiding obstacles.

7. **Edge Detection:** It is an image processing technique for finding the boundaries of objects within images. It works by detecting discontinuities in brightness. Edge detection is used for image segmentation and data extraction in areas such as image processing, computer vision (CV), and machine vision.

8. **Travelling Salesman Problem (TSP):** It describes a salesman who must travel between N cities. The order in which he does so is something he does not care about, as long as he visits each once during his trip, and finishes where he was at first.

KEYWORDS

- ant colony optimization
- artificial intelligence
- edge detection
- image processing
- metaheuristic
- path planning
- robotic navigation
- swarm intelligence
- traveling salesman problem

REFERENCES

1. Dorigo, M., (1992). *Optimization, Learning and Natural Algorithms*. PhD dissertation, Department of Electronics, Politecnico di Milano, Milan, Italy.
2. Gutowitz, H., (1994). Complexity-seeking ants. *Third European Conference on Artificial Life* (pp. 1–16). Brussels.
3. Dorigo, M., & Gambardella, L. M., (1997). Ant colony system: A cooperative learning approach to the traveling salesman problem. *IEEE Transactions on Evolutionary Computing, 1*(1), 53–66.
4. Eberhart, R., Shi, Y., & Kennedy, J., (2001). Swarm intelligence. *Sixth International Symposium on Micro Machine and Human Science* (pp. 223–236). Nagoya, Japan.
5. Bonabeau, E., Dorigo, M., & Theraulaz, G., (2003). *Swarm Intelligence: From Natural to Artificial Systems.* Emerging technology conference. Santa Clara, CA.
6. Louis, R., David, B., & Niccolo, P., (2006). *Crowds vs. Swarms, a Comparison of Intelligence.* Swarm/human blended intelligence workshop (SHBI), Cleveland, OH, IEEE Xplore.
7. Rai, R., (2009). Optimization of autonomous multi-robot path planning and navigation using swarm intelligence. *National Conferences on Lean Manufacturing Implementations: The Future of Process Industries (Leman'2009)*. Green Hills Engineering College, Shimla.
8. Guowei, Y., & Fengchang, X., (2011). Research and analysis of image edge detection algorithm based on the MATLAB. *Procedia Engineering, 15*, 1313–1318.
9. Rai, R., (2011). A hybrid framework for robot path planning and navigation using ACO & Dijkstra. *IJCA Proceedings on International Symposium on Devices MEMS, Intelligent Systems and Communication (ISDMISC)* (pp. 19–24). Published by Foundation of Computer Science, New York, USA.

10. Rai, R., (2013). Ant-based swarm computing techniques for edge detection of images: A brief survey. In: *International Journal of Emerging Technology and Advanced Engineering* (Vol. 3, No. 4/43). ISSN: 2250-2459, ISO 9001:2008 certified journal.
11. Rai, R., (2013). AASC: Advanced ant-based swarm computing for detection of edges in imagery. *International Journal of Emerging Technology and Advanced Engineering (IJETAE)* (Vol. 3, No. 12). ISSN: 2250-2459, ISO: 9001:2008 Certified Journal.
12. Abubakar, M. A., & Eleyan, A., (2015). A multi-resolution approach for edge detection using ant colony optimization. In: *23rd Signal Processing and Communications Applications Conference (SIU)* (pp. 1580–1584). Malatya, Turkey.
13. Jahanzaib, S., & Tarique, A., (2015). Artificial intelligence and its role in near future. *Journal of Latex Class Files, 14*(8), 1–11.
14. Dorrani, Z., & Mahmoodi, M. S., (2016). Noisy images edge detection: Ant colony optimization algorithm. *Journal of AI and Data Mining, 4*(1), 77–83.
15. Belal, A., Shivank, S. C., Subham, B., Gayathri, P., & Santhi, H., (2017). Analysis of traveling salesman problem. *IOP Conf. Series: Materials Science and Engineering, 263*, 042085.
16. Beni, G., & Wang, J. (1993). Swarm Intelligence in Cellular Robotic Systems. *Proceed. NATO Advanced Workshop on Robots and Biological Systems, Tuscany, Italy, June 26–30 (1989).* Berlin, Heidelberg: Springer. pp. 703–712. doi:10.1007/978-3-642-58069-7_38. ISBN 978-3-642-63461-1.
17. Eberhart, R. C., & Kennedy, J. (1995). A new optimizer using particle swarm theory. In Proceedings of the sixth international symposium on micro machine and human science (pp. 39–43), Nagoya, Japan. Piscataway: IEEE.
18. Kazemian, M., Ramezani, Y., Lucas, C., & Moshiri, B., (2006). Swarm Clustering Based on Flowers Pollination by Artificial Bees, Swarm Intelligence in Data Mining, *34,* 191–202.
19. Karlson, P., & Lüscher, M. 'Pheromones': a New Term for a Class of Biologically Active Substances. Nature 183, 55–56 (1959). https://doi.org/10.1038/183055a0.

CHAPTER 2

USER VS. SELF-TUNING OPTIMIZATION: A CASE STUDY ON IMAGE REGISTRATION

JOSE SANTAMARIA LOPEZ and MARIA L. RIVERO-CEJUDO

Department of Computer Science, University of Jaen, Spain

E-mail: jslopez@ujaen.es (J. S. Lopez)

ABSTRACT

In the last few years, there has been an increased interest in providing new soft computing (SC) algorithms, e.g., evolutionary algorithms (EAs), in which it is not needed the tuning of the control parameters. Usually, this new approach is named self-tuned and several models have been proposed to date. In our study, we aim to, firstly, describe how self-tuning EAs work and, secondly, provide a computational comparison of algorithms dealing with a specific computer vision (CV) task, named image registration (IR), as baseline. In particular, the development of automated IR methods is a well-known issue within the CS field and it was largely addressed from multiple viewpoints. IR has been applied to a high number of real-world scenarios ranging from remote sensing to medical imaging, artificial vision, and computer-aided design. Then, in this contribution it is provided a comprehensive analysis of the self-tuning approach by means of the experimental comparison of several EA-based IR algorithms proposed in the state-of-the-art.

2.1 INTRODUCTION

Soft computing (SC) [1] is a term applied to a field within computer science that is characterized by the use of inexact solutions to computationally-hard tasks such as the NP-complete problems, in which a global optima solution cannot be derived in (desirable) polynomial time. In particular, SC differs from conventional (hard) computing in that, unlike hard computing, it is tolerant regarding to imprecision, uncertainty, partial truth, and approximation. Then, the guiding principle of SC is exploiting the tolerance for such an imprecision, uncertainty, partial truth, and approximation in order to achieve tractability, robustness, and low solution cost. The main SC paradigms include fuzzy systems (FS), evolutionary computation (EC), and artificial neural computing (ANC), among others.

Specifically, EC is based on the principles of biological evolution. The precise origin of EC is difficult to determine. A number of authors are commonly cited for originating EC [2]. In the last decades, EC has demonstrated its ability to deal with complex real-world problems [3]. Nowadays, three main different forms of EC are well-known: evolutionary programming (EP), evolution strategies (ESs), and genetic algorithms (GAs). These kinds of EC algorithms are also known as evolutionary algorithms (EAs). Additionally, several further approaches have emerged that adopt other mechanisms from nature. For instance, ant colony optimization (ACO) [4] mimics the behavior of a colony of ants finding the shortest path from the nest to a location containing food. Moreover, other general-purpose optimization algorithms known as metaheuristics are also considered in this category.

Similarly, particle swarm optimization (PSO) [5, 6] and artificial immune system (AIS) [7] are recent contributions in the field, among others. The former approach is motivated by the social behavior of organisms such as bird flocking and fish schooling, and the latter is based on principles of immune system defenses against viruses. The reader can deepen on the specific literature of EC in [8].

On the other hand, it is well-known the necessity for accurately tuning of the algorithm-dependent parameters, i.e., the control parameters, of any EA before testing its performance when dealing with any non-linear optimization problem. Thus, it has been proven the performance of any algorithm will largely depend on the accurate setting of these parameters. The better the user fixes their values, the best the final optimization result will be. A trend that has emerged recently is to make these control parameters automatically adapt to the problem, thereby liberating the user from the tedious and time-consuming task of manual setting. This recent approach is known as either *Self-tuned* or *Self-adapted* EAs which have been successfully applied in real-world problems from industrial control systems [9, 10], cloud computing [11], and other fields [12].

In the last two decades, the EC paradigm has demonstrated its ability to deal with complex real-world problems within computer vision (CV) field. As an example, several special issues and books on the topic have been published in international forums in the last few years [13, 14]. Specifically, EC has been successfully applied to tackle the image registration (IR) problem, which is one of the most important operations of image processing systems. To sum up, from an optimization viewpoint, IR involves finding the optimal *transformation* achieving the best fitting between, usually, two images, named *scene* and *model*. Recent contributions reviewing the state-of-the-art (SOTA) on IR methods based on EAs can be found in [15].

In this contribution, we aimed at providing a practical review of the latest developments in IR using this kind of optimization algorithms. Specifically, our research is an original computational study in which several user and self-tuning IR methods have been compared when dealing with 3D reconstruction problems of range images. Among these methods, we considered some of the most recent proposals using a self-tuned approach. To do so, several range image datasets from the signal analysis and machine perception laboratory (SAMPL) at the Ohio State University have been used.

The structure of this chapter is as follows. Section 2.2 introduces the problem of IR and describes the most significant strengths and drawbacks of the early proposals. Subsequently, Subsections 2.2.1 and 2.2.2 provide a brief review of the SoTA and the description of those contributions based on the Self-tuning approach. Next, Section 2.3 is devoted to provide the reader with a practical evaluation of these new IR methods. In particular, several EA-based IR methods have been included in the experimental section tackling with instances of range IR problems. Finally, a concluding analysis of the reported results is accordingly introduced in Section 2.4.

2.2 IMAGE REGISTRATION (IR) AND EVOLUTIONARY COMPUTATION (EC)

It has been proven that there is not a universal design for a hypothetical IR method that could be applicable to every real-world application [16]. However, IR methods consist of the following four components:

- Two input images named as *scene* $I_s = \{p_1, p_2, ..., p_n\}$ and *model* $I_m = \{\bar{p}'_1, \bar{p}'_2, \cdots, \bar{p}'_m\}$, with p_i and p_j^0 being image points.
- A *registration transformation*, named *f*: It corresponds to a geometric/parametric function that relates the coordinate systems of both images.
- A *similarity metric function*, named *F*. It is aimed at assessing the level of resemblance (i.e., the degree of overlapping) between the transformed scene image (i.e., $f^0(I_s)$) and the model one.
- An *optimization procedure*: The optimizer regards to a method seeking for the optimal registration transformation (*f*). Likewise, it is necessary to provide a search space for the suitable representation of the IR solutions.

As stated, due to the non-linear nature of the inherent optimization procedure of the IR problem, an iterative procedure is often followed. As depicted in Figure 2.1, such procedure usually stops once convergence is achieved, specifically, after the *F* value (i.e., the similarity metric) is lower than a given tolerance threshold. Then, we focused our attention on the optimization task, because of its vital importance in the success of the whole performance of the IR method. Moreover, two different search space schemes have been adopted in the SoTA: (a) the *matching-based* approach, in which the optimization process is aimed at looking for an optimal set of pairs of similar image points; and (b) the *parameter-based* one, which directly explores the search space of the values of IR solutions, i.e., *f*.

The early proposals tackling the IR problem were based on the first approach (i.e., the *matching-based*). This is the case of the well-known ICP method [17]. Its main pitfall is its sensitiveness to the initial IR transformation (*f*) given by the user [18]. Then, ICP is prone to get trapped in local optima solutions.

A detailed description of the IR framework is out of the scope of this contribution. We refer the interested reader to [16]. Likewise, the formulation of the IR problem is dependent on the particular environment in which it is involved (remote sensing, medical imaging, CAD, etc.). Nevertheless, the subsequent computational study is based on addressing 3D reconstruction problems considering range images as faced in previous contributions [15].

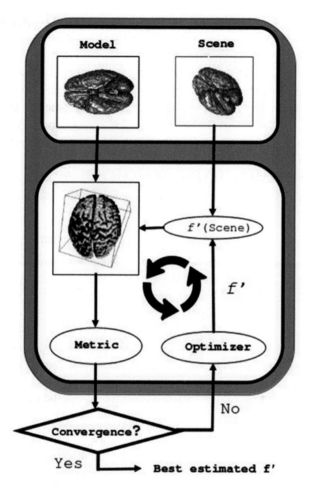

FIGURE 2.1 The flow-diagram of the IR optimization process.

The latter concern regarding ICP has been the main goal to be addressed by those recent contributions about IR using solutions based on EC, due to they successfully work without requiring a good initial estimation of the image alignment. That advantage is mainly motivated by the global optimization nature of the evolutionary approaches, which allows them to perform a robust search in complex and ill-defined search spaces. Since the first attempts facing the IR problem using GAs [19], this topic has become a very active niche of research and a significant number of proposals have been contributed to date.

A deep survey on IR and EC is out of the scope of this work. Nevertheless, next we introduce a selection of, in our opinion, those most relevant contributions in the field. More importantly, first, we consider those IR methods using a user-based tuning approach, and next, the remaining revised contributions regard to those based on the self-tuned approach. Several of them have been considered in the subsequent experimental section.

2.2.1 USER-BASED TUNING APPROACHES

2.2.1.1 GARCÍA-TORRES ET AL.'S METHODS

In 2014 [32], the authors contributed with a computational study on the application of cutting-edge population-based optimization algorithms to the problem of IR. Specifically, the artificial bee colony (ABC), harmony search (HS), and biography-based optimization (BBO) algorithms were adapted for addressing the IR problem of range images. Briefly, ABC is a new swarming variant inspired in the population-based approach, which is based on the intelligent behavior of honeybee swarms. The algorithm works just by including a common area in the hive, so-called the dancing area, where bees share and exchange information about food sources. Bees identify the quality of food source by means of the duration of dancing which is determined by the nectar contained in the food source being exploited and its distance to the hive. On the other hand, HS is based on analogies with the natural phenomena of the musical process. It is focused on searching for the perfect state of harmony, an analogy to the optimization process. Last, BBO is based on mathematical models of biography describing natural ways of distributing species, i.e., how species migrate, how they arise and become extinct. In the conducted experiments, the reported results shown the promising outcomes achieved by these three recent optimization techniques tackling the IR problem.

2.2.1.2 MA ET AL.'S METHOD

Ma et al. [20] proposed an intensity-based IR method for remote sensing. This IR variant made use of the hybridization of both the orthogonal learning (OL) and the DE algorithms (OLDE). It has been demonstrated that the OL strategy provides an efficient alternative when searching for in complex problem spaces. In OLDE, the OL strategy guides the DE algorithm to select promising search directions towards the global optimum. In particular, they provide a crossover component for DE based on OL, named *Orthogonal Crossover*. They tested the performance of OLDE against the one from other EA-based IR methods. The experimental section considered 2D images of RADARSAT SAR sensors. The reported results showed that OLDE was the best IR algorithm which achieved robust and efficient outcomes with respect to the initial values of the IR transformation.

2.2.1.3 YANG ET AL.'S METHOD

In 2015 [21], the authors tackled non-rigid multi-modal IR scenarios using the NMI metric and proposed a hybrid method that combines both the L-BFGS-B and the cat swarm optimization (CSO) algorithms, named HLCSO. On the one hand, CSO makes use of cooperative co-evolving and block grouping as the grouping strategy to capture the interdependency among variables. On the other hand, L-BFGS-B provides a way of achieving faster convergence and higher accuracy of the final solution. Moreover, the roulette wheel method is introduced into both the seeking and the tracing modes of HLCSO in order to improve the performance of the algorithm. Extensive experiments on 3D CT, PET, T1, T2 and PD weighted MR images demonstrated the superiority of the proposed hybrid approach, HLCSO, against separately using the L-BFGS-B and the CSO methods.

2.2.1.4 PANDA ET AL.'S METHOD

In this contribution [22], the authors introduced a novel evolutionary rigid-body docking (ERBD) algorithm for addressing medical IR problems. They considered different image data formats like MR, CTs, and PET images of the brain of the same patient, all of them from the well-known retrospective image registration evaluation (RIRE) project. Regarding their proposal, i.e., ERBD, the docking process is successfully used in the design of drugs and it has never been used in IR. In particular, the docking task predicts the optimal configuration and energy between the protein and ligand. It changes the orientation of molecules and maximizes their contact. When the interaction energy between the protein and the ligand is minimized, they achieve best binding. Minimization of the energy also preserves the discontinuities in the low texture region. In their contribution, the authors used a canonical GA within ERBD to minimize the total energy. From the reported results, it is showed that the proposed method leads to higher IR accuracy since the interaction of both the energy and the MI were considered as the similarity metrics.

2.2.2 SELF-TUNING APPROACHES

2.2.2.1 SANTAMARÍA ET AL.'S METHOD

This work [23] presented a new self-tuned EA model to deal with IR problems facing pair-wise range IR problem instances. The specific design of their self-adaptive optimization method, named *SaEvO*, takes advantage of the synergy between two different EAs: a memetic algorithm (MA) [24] based on differential evolution (DE) [25] and the variable neighborhood search (VNS) [26], and combined with an AIS algorithm [7]. Actually, SaEvO represents a design improvement with respect to the method proposed in Ref. [33]. In a similar manner to DE, VNS is easy to implement and it requires few control parameters. Moreover, VNS extends the capabilities of DE by means of carrying out a local search step. While the optimization stage of MA (first stage of SaEvO) is aimed at searching for IR solutions, the latter stage (based on AIS) is focused on tuning the control parameters, i.e., for both the DE and the VNS algorithms. The conducted experiments considered several range image datasets obtained from the well-known public repositories, e.g., the SAMPL, and others from a Konica Minolta c Laser Range Scanner. Their conducted experiments reported outstanding results when comparing SaEvO against several IR algorithms in the SoTA.

2.2.2.2 DE FALCO ET AL.'S METHOD

De Falco et al. [27] contributed with an adaptive invasion-based model (AIM) to tackle range IR problems. Their AIM-dDE method made use of a distributed DE (dDE) algorithm which considers a migration model. The latter is inspired by the natural phenomenon known as *biological invasion*. Briefly, at each generation t, the subpopulation $P_p(t)$ of each node p performs a sequential DE until t equals t_{max} generations, and at every T generations, neighboring subpopulations exchange individuals. Then, at a given invasion time, $t = kT$, the pool of individuals in each subpopulation $P_p(t)$ sends to its *propagule* (i.e., neighbors) according to $M_{Pp(t)}$ and it is carried out by gathering the individuals of $P_p(t)$ which are better than its average fitness. Next, a *founding subpopulation* formed by both native and exogenous individuals is built in each node by means of adding this huge number of invading individuals to the local subpopulation. In this manner, the authors provided a novel mechanism based on the founding subpopulation as a source of diversity exploitable by their algorithm to properly enhance its capability to search for new niches of solutions. Moreover, their AIM-dDE made use of an adaptive strategy for searching for the optimum control parameters, i.e., named *RandAvg* and *ChAvg*. The authors collected several range image datasets from a well-known source in the field, i.e., the SAMPL. They provided parallel versions of the proposed algorithms by making use of the message passing interface (MPI) paradigm. Their proposed range IR method was compared against several algorithms in the SoTA and the reported results suggested that the *AIM-dDE-ChAvg* variant turn out to be the best one in all the considered tests.

2.2.2.3 LI ET AL.'S METHOD

Li et al. [28] proposed a design model-based inspection algorithm to tackle with IR problems of range images. This method combined the DE algorithm and an EFT-based 3D point descriptor to improve the IR results. This point property-based DE algorithm was named PDE, in which the control parameters F_i and CR_i were set adaptively as proposed in Ref. [29]. Initially, the contributed algorithm randomly selects three-point pairs n ($n < n_p$) times to form the candidate solution pool (CSP). The solutions in CSP are approximate and provide pre-registration for the two tested image models. In practice, with the increasing complexity of the geometry of the image, the rate of false solutions becomes higher. Thus, PDE generates the population from two different sets: the X_1 set, is randomly generated, and the X_2 set is generated based on the CSP. Additionally, their design made use of elitism in which the k best individuals of X_2 are considered to replace the k worst individuals of X_1. This strategy was aimed at speeding up the convergence of the algorithm. Moreover, the mutation strategy proposed in Ref. [29] was also included in PDE. The authors conducted several IR tests, and it was considered both synthetic and range image datasets. PDE was compared against three deterministic (non-coarse) algorithms and other five heuristic algorithms. Finally, the authors concluded that PDE was insensitive (robust) to outliers, and PDE can obtain the global optima solutions successfully for all tested situations.

2.2.2.4 COCIANU AND STAN'S METHOD

In 2019 [30], the authors proposed a new IR method based on the hybridization of a self-adaptive ES algorithm and an accelerated PSO algorithm (ES-APSO). This contribution tackles with signature recognition problems. In particular, their PSO technique is based on the firefly optimization (FFO) [31] algorithm. FFO is based on the flashing behavior of fireflies. First, fireflies are unisex, so that one firefly will be attracted to other fireflies regardless of their sex. The second rule states that the attractiveness is proportional to the brightness, and both fireflies decrease as their distance increases. The third rule concerns to the brightness of a firefly, which is determined by the landscape of the objective function (i.e., F, see Section 2.2). ES-APSO was tested by considering several 2D images representing signatures, and it was compared against several canonical versions of FFO. The reported experimental results revealed the suitable application of ES-APSO.

2.3 A CASE STUDY

2.3.1 EXPERIMENTAL SETUP

This section is aimed at presenting a number of experiments to study the performance of the five analyzed user and self-tuned optimization algorithms facing IR problems of range images, named RIR. Specifically, we considered a benchmark suite for the following IR methods of the SoTA:

- **User-Based Tuning:** ABC, HS, and BBO [32];
- **Self-Tuning:** SaEvO [23] and StEvO [33].

All the five considered algorithms were implemented in C++ and compiled with the GNU/g++ tool. We adapted all the tested methods by using both the same representation of the rigid transformation (f) and the objective function (i.e., F, see Section 2.2) in Ref. [23] in order to accomplish a fair comparison.

Finally, all the tested methods were run on an Intel Pentium IV 2.6 MHz PC and 2GB RAM. We considered the values of the control parameters of each of the three user-based tuning methods (i.e., ABC, HS, and BBO) as those used in their original contribution [32].

2.3.2 DATASETS AND PARAMETER SETTINGS

In order to ease the comparison with the results reported in other contributions in the field [34, 35], our results correspond to a number of pair-wise RIR problem instances using different range datasets obtained from the well-known public repository of the *Signal Analysis and Machine Perception Lab* (SAMPL). Specifically, we used six range datasets, named in previous contributions [35]: "Frog," "Bird," "Tele," "Lobster," "Angel," and "Dock."

Besides, we have defined several pair-wise RIR problem scenarios using different overlapping degrees between pairs of adjacent images. Specifically, four, and six RIR problem instances were considered using pairs of range images of the SAMPL's datasets at 20 and 40 rotation degrees of misalignment between the adjacent views. Then, we designed two different RIR problem scenarios regarding the rotation degree, i.e., 20 and 40, from which ten different problem instances have been generated.

In order to avoid execution dependence, 30 different runs have been performed for each of the five tested RIR algorithms when facing each of the four problem scenarios, i.e., considering 20 and 40° of image overlapping. Moreover, all the tested algorithms start from an initial population of random solutions. Moreover, we considered the CPU time as the stop criterion in order to perform a fair comparison between the methods included in this study. Different time limits were tested, and 20 seconds was determined as a good threshold allowing all the methods to achieve accurate outcomes.

2.3.3 ANALYSIS OF RESULTS

Tables 2.1 and 2.2 show the statistical outcomes regarding the 30 different runs carried out by each of the five RIR algorithms facing the two RIR problem scenarios (i.e., 20 and 40° of overlapping). In particular, each column of the former tables refers to the range dataset, the method, and the minimum, maximum, median, and standard deviation values of the F function (see Section 2.2) in those 30 runs. The unit length is always squared millimeters. Two different algorithm families are distinguished, that of the three user-based tuning methods: ABC, BBO, and HS; and that of the two self-tuning RIR methods: StEvO and SaEvO. The algorithm with the best (lowest) minimum and median results is accordingly highlighted using bold-font as well as the overall best median value is underlined. Regarding Figure 2.2, the bar-graph depicts the accumulated median error (AME) for each of the five RIR methods when dealing with each of the ten problem scenarios of RIR.

It can be shown in Tables 2.1 and 2.2 and that the two self-tuned EC-based RIR methods, StEvO, and SaEvO, achieve the best score regarding accuracy, i.e., the lowest minimum value of F, and none of the other three (ABC, BBO, and HS) outperform the performance of any of them. Specifically, SaEvO achieves better accurate results than StEvO when dealing with the most complex scenario, i.e., range images acquired when rotating the turning table 40°.

TABLE 2.1 Outcomes Regarding the 20 RIR Problem Scenario

Dataset	Method	Min.	Max.	Median	Std. dev.
Angel	ABC	0.2470	0.5289	0.5265	0.1007
	BBO	0.2576	0.9554	0.5361	0.2692
	HS	0.2494	0.9535	0.4852	0.2724
	StEvO	0.2446	0.5266	0.4726	0.0884
	SaEvO	0.2446	0.9440	0.4715	0.2119
	ABC	0.1167	0.9009	0.14388	0.2451
	BBO	0.1263	0.9301	0.4096	0.2759
Bird	HS	0.1170	0.9188	0.8989	0.3603
	StEvO	0.1123	0.5977	0.1123	0.1569
	SaEvO	0.1123	0.5997	0.1123	0.1571
	ABC	0.1226	0.7733	0.1282	0.1798
	BBO	0.1649	0.8690	0.8690	0.2023

TABLE 2.1 *(Continued)*

Dataset	Method	Min.	Max.	Median	Std. dev.
Frog	HS	0.1260	0.8751	0.3460	0.2749
	StEvO	0.1187	0.5386	0.1227	0.1354
	SaEvO	0.1189	0.5296	0.1225	0.1337
	ABC	0.0752	0.8691	0.1011	0.1817
	BBO	0.0829	0.8699	0.6035	0.2292
Tele	HS	0.0754	0.8721	0.7873	0.2963
	StEvO	0.0734	0.1072	0.0750	0.0080
	SaEvO	0.0734	0.1070	0.0752	0.0080

Also, this outstanding behavior of SaEvO is shown when considering the median value, which reveals the significant robustness of the method due to it achieves the best score in 8 of the 10 addressed scenarios. Moreover, Figure 2.2 highlights the overall performance of the tested RIR methods and both SaEvO and StEvO achieve outstanding results. Finally, it is remarkable the promising performance of ABC (see Figure 2.2) compared with the two best self-tuned approaches, which suggests that there is room for improvement for the former method.

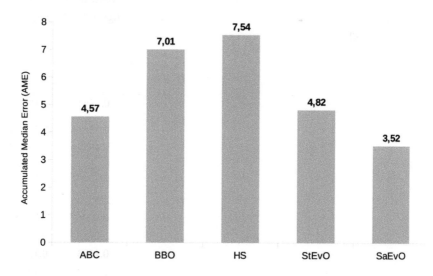

FIGURE 2.2 Bar-graph depicting the AME value for each of the tested RIR methods.

2.4 CONCLUDING REMARKS

In the last few years, there has been an increased interest in providing new optimization approaches based on EAs for tackling CV tasks, specifically those focused on IR. Among those approaches, novel strategies have risen in which it is not needed the tuning of the control parameters, and they have been named Self-tuned EAs.

In this work, we aimed to provide a brief review of the SoTA on IR and several models using the self-tuning approach have been introduced. Next, a computational study focused on the comparison of the performance of several of these IR methods dealing with 3D reconstruction problems of range images have been carried out. The statistical results reported in this work has shown the outstanding performance achieved by those IR methods using the self-tuning model of optimization. Moreover, it has been highlighted the promising outcomes provided by other EAs which reveals that there exists room for improvement.

TABLE 2.2 Outcomes Regarding the 40 RIR Problem Scenario

Dataset	Method	Min.	Max.	Median	Std. Dev.
Angel	ABC	0.3542	0.9098	0.8837	0.2210
	BBO	0.3690	0.9674	0.8187	0.2435
	HS	0.3553	0.9567	0.9567	0.2665
	StEVO	**0.3491**	0.8526	**0.8072**	0.2025
	SaEvO	0.3493	0.9439	0.8083	0.2175
	ABC	0.2124	0.9308	0.8299	0.2829
	BBO	0.2426	0.9419	0.9338	0.2189
Bird	HS	0.2165	0.9430	0.8966	0.3058
	StEvO	0.2040	0.9259	0.9259	0.2807
	SaEvO	**0.2028**	0.9268	**0.4542**	0.3051
	ABC	0.2717	0.8410	0.6489	0.2015
	BBO	0.4794	0.9191	0.8990	0.0823
Frog	HS	0.4026	0.9005	0.8597	0.1161
	StEvO	**0.2507**	0.7702	0.7691	0.2056
	SaEvO	0.2535	0.7724	**0.4643**	0.1963
	ABC	0.1082	0.8607	0.3120	0.2222
	BBO	0.1198	0.8837	0.3897	0.2300

TABLE 2.2 *(Continued)*

Dataset	Method	Min.	Max.	Median	Std. Dev.
Tele	HS	0.1095	0.9222	0.8465	0.3072
	StEvO	0.1054	0.4706	0.3059	0.1222
	SaEvO	**0.1049**	0.8062	**0.3052**	0.1667
	ABC	0.1755	0.8774	0.2934	0.2226
	BBO	0.3203	0.9438	0.9065	0.1789
Dock	HS	0.2385	0.9480	0.8576	0.1944
	StEvO	0.1588	0.8525	0.8525	0.1878
	SaEvO	**0.1585**	0.4821	**0.1681**	0.1134
	ABC	0.2745	0.8220	0.7054	0.1530
	BBO	0.3249	0.9406	0.6898	0.1555
Lobster	HS	0.2665	0.9257	0.6089	0.1964
	StEvO	0.2512	0.8450	0.8450	0.1945
	SaEvO	**0.2504**	0.7582	0.5389	0.1894

KEYWORDS

- adaptive invasion-based model
- ant colony optimization
- artificial bee colony
- artificial immune system
- artificial neural computing
- biography-based optimization
- candidate solution pool

REFERENCES

1. Zadeh, L., (1994). Soft computing and fuzzy logic. *IEEE Software, 11*(6), 48–56.
2. De Jong, K., (2002). *Evolutionary Computation*. The MIT Press.
3. Glover, F., Laguna, M., & Mart´ı, R., (2003). Scatter search. In: Ghosh, A., & Tsutsui, S., (eds.), *Advances in Evolutionary Computation: Theory and Applications* (pp. 519–537). New York: Springer-Verlag.

4. Dorigo, M., & Di Caro, G., (1999). The ant colony optimization meta-heuristic. In: Corne, D., Dorigo, M., & Glover, F., (eds.), *New Ideas in Optimization* (pp. 11–32). Nueva York, NY, EE. UU.: McGraw-Hill.
5. Clerc, M., (2006). *Particle Swarm Optimization*. ISTE Publishing Company.
6. Kennedy, J., & Eberhart, R., (2001). *Swarm Intelligence*. Morgan Kaufmann, San Francisco, CA.
7. Castro, E. D., & Timmis, J., (2002). *Artificial Immune Systems: A New Computational Intelligence Approach*. Springer, Berlin.
8. Zelinka, I., (2015). A survey on evolutionary algorithms dynamics and its complexity-mutual relations, past, present and future. *Swarm and Evolutionary Computation, 25*, 2–14.
9. Sharkawy, A., (2010). Genetic fuzzy self-tuning PID controllers for antilock braking systems. *Engineering Applications of Artificial Intelligence, 23*(7), 1041–1052.
10. Chopra, V., Singla, S., & Dewan, L., (2014). Comparative analysis of tuning a PID controller using intelligent methods. *Acta Polytechnica Hungarica, 11*, 235–249.
11. Jeyarani, R., Nagaveni, N., & Ram, R. V., (2011). Self-adaptive particle swarm optimization for efficient virtual machine provisioning in cloud. *International Journal of Intelligent Information Technologies, 7*(2), 25–44.
12. Harrison, K., Engelbrecht, A., & Ombuki-Berman, B., (2018). Self-adaptive particle swarm optimization: A review and analysis of convergence. *Swarm Intelligence, 12*(3), 187–226.
13. Nachtegael, M., Kerre, E., Damas, S., & Van, D. W. D., (2009). Special issue on recent advances in soft computing in image processing. *Int. J. Approx. Reason., 50*(1), 1–2.
14. Olague, G., (2016). *Evolutionary Computer Vision*. Springer, Berlin.
15. Santamaría, J., Cord´on, O., & Damas, S., (2011). A comparative study of state-of-the-art evolutionary image registration methods for 3D modeling. *Comput. Vis. Image Underst., 115*, 1340–1354.
16. Zitova, B., & Flusser, J., (2003). Image registration methods: A survey. *Image Vision Comput., 21*, 977–1000.
17. Besl, P. J., & McKay, N. D., (1992). A method for registration of 3D shapes. *IEEE T. Pattern Anal. Mach. Intell., 14*, 239–256.
18. Rusinkiewicz, S., & Levoy, M., (2001). Efficient Variants of the ICP algorithm. In: *Third International Conference on 3D Digital Imaging and Modeling (3DIM'01)* (pp. 145–152). (Quebec, Canada).
19. Fitzpatrick, J., Grefenstette, J., & Gucht, D., (1984). Image registration by genetic search. In: *IEEE Southeast Conference* (pp. 460–464). Louisville, EEUU.
20. Ma, W., Fan, X., Wu, Y., & Jiao, L., (2014). An orthogonal learning differential evolution algorithm for remote sensing image registration. *Mathematical Problems in Engineering*.
21. Yang, F., Ding, M., Zhang, X., Hou, W., & Zhong, C., (2015). Nonrigid multi-modal medical image registration by combining L-BFGS-B with cat swarm optimization. *Information Sciences, 316*, 440–456.
22. Panda, R., Agrawal, S., Sahoo, M., & Nayak, R., (2017). A novel evolutionary rigid-body docking algorithm for medical image registration. *Swarm and Evolutionary Computation, 33*, 108–118.

23. Santamaría, J., Damas, S., Cord´on, O., & Escamez, A., (2013). Self-adaptive evolution toward new parameter free image registration methods. *IEEE Transactions on Evolutionary Computation, 17*, 545–557.
24. Ong, Y., Lim, M. H., & Chen, X., (2010). Memetic computation past, present and future. *IEEE Computational Intelligence Magazine, 5*, 24–31.
25. Storn, R., (1997). Differential evolution: A simple and efficient heuristic for global optimization over continuous spaces. *J. Global Optim.*, 341–359.
26. Mladenovic, N., & Hansen, P., (1997). Variable neighborhood search. *Comput. Oper. Res., 24*, 1097–1100.
27. Falco, I. D., Cioppa, A. D., Maisto, D., Scafuri, U., & Tarantino, E., (2014). Using an adaptive invasion-based model for fast range image registration. In: *GECCO'14: Proceedings of the 2014 Genetic and Evolutionary Computation Conference* (pp. 1095–1102). Vancouver, Canada.
28. Li, T., Pan, Q., Gao, L., & Li, P., (2017). Differential evolution algorithm-based range image registration for free-form surface parts quality inspection. *Swarm and Evolutionary Computation, 36*, 106–123.
29. Zhang, J., Member, S., & Sanderson, A., (2009). Jade: Adaptive differential evolution with optional external archive. *IEEE Transactions on Evolutionary Computation, 13*(5), 1–14.
30. Cocianu, C., & Stan, A., (2019). New evolutionary-based techniques for image registration. *Applied Sciences, 9*(176), 176.
31. Yang, X., (2014). *Nature-Inspired Optimization Algorithms*. Elsevier.
32. García-Torres, J. M., Damas, S., Cord´on, O., & Santamaría, J., (2014). A case study of innovative population-based algorithms in 3D modeling: Artificial bee colony, biogeography-based optimization, harmony search. *Expert Syst. Appl., 41*(4), 1750–1762.
33. Santamaría, J., Damas, S., Garcia-Torres, J. M., & Cord´on, O., (2012). Self-adaptive evolutionary image registration using differential evolution and artificial immune systems. *Pattern Recognition Letters*.
34. Salvi, J., Matabosch, C., Fofi, D., & Forest, J., (2007). A review of recent range image registration methods with accuracy evaluation. *Image Vision Comput., 25*(5), 578–596.
35. Silva, L., Bellon, O. R. P., & Boyer, K. L., (2005). Precision range image registration using a robust surface interpenetration measure and enhanced genetic algorithms. *IEEE T. Pattern Anal. Mach. Intell., 27*(5), 762–776.

CHAPTER 3

SECURE COMMUNICATION USING A NOVEL 4-D DOUBLE SCROLL CHAOTIC SYSTEM

PUSHALI TRIKHA and LONE SETH JAHANZAIB

Department of Mathematics, Jamia Millia Islamia, New Delhi, India, E-mail: lone.jahanzaib555@gmail.com (L. S. Jahanzaib)

ABSTRACT

In this chapter, a novel 4-D chaotic system is constructed and analyzed by means of time series; phase portrait; existence and uniqueness of solution; dissipative and symmetric character; equilibrium point analysis; lyapunov exponent; bifurcation diagram, etc. The novel system is synchronized using a novel synchronization technique viz. multi-switching phase synchronization and its application in secure communication is illustrated in the field of secure communication.

3.1 INTRODUCTION

A chaotic system is a dynamical system with at least one positive L.E. (lyapunov exponent), with one zero L.E. and with one negative L.E. so as to have the bounded solution. The smallest dimension for a continuous dynamical system to be a chaotic system is three. Chaotic systems because of their high sensitivity to I.C. (initial conditions) and parametric values find application across various disciplines such as secure communication [1], image encryption [2], control systems, weather, and earth disciplines [3], circuits, etc., using novel synchronization methods such as compound difference [4], combination difference [5], double compound combination

[6], dual combination anti-synchronization [7], etc. Some classic examples of chaotic systems are Lorenz system [8, 9], Chen system, Lu system, etc. Many new non-standard chaotic and hyperchaotic systems have been developed recently, such as in Refs. [10–16] with their ability of not being easily guessed by the intruders. Many techniques exist to analyze the chaotic dynamical systems-such as studying about their parameter values, studying their rate of divergence from nearby initial points, studying the intersection of orbits with subspace, studying the system properties along solutions that does not change with time, i.e., about their equilibrium points, studying the entropy character, studying their hidden or self-excited attractors [17–19]. The type of equilibrium points viz. stable or unstable, hyperbolic or non-hyperbolic and number of points such as zero, finite, countably infinite, uncountably infinite lying along a line, surface, etc., are also of importance in unveiling the hidden dynamics of chaotic systems.

Though chaotic systems that may seem slightest of different from each other are capable of showing entirely different dynamics. In such a situation, making one system follow the path of other asymptotically involves designing suitable controllers-choosing carefully the desired synchronization type using some control methods. Nowadays, many robust synchronization methods are available depending on whether the parameters are determined or undetermined, whether any surface is involved while studying the dynamics of the systems, etc. Also, to increase the slightest chances of being hacked, the type of synchronization to be performed is made complex by clubbing two or more synchronization techniques.

Motivated by the above work, we have proposed a novel 4-D chaotic system with three equilibrium points (one stable and two unstable). The proposed system exhibits double scroll attractor. We have studied the introduced system through LEs [20], symmetry, dissipative, Kaplan Yorke dimension, Poincare map, bifurcation diagram, etc. These findings deepen the understanding of the diverseness of the dynamics in chaotic system [21–23] and adore the chaos theory [24, 25]. Also, the novel system has been synchronized with its identical system in the presence of system uncertainties using a novel synchronization technique viz. multi-switching phase synchronization. The application of the above technique has been illustrated in secure communication. The obtained results clearly show the efficacy of the synchronization method in this area. Designing such nonstandard chaotic systems helps improve the security of transmission of information as these remain unknown to the hackers. Also, since the chaotic systems are highly sensitive to I.C. and parameter values, hacking the exact details is very difficult for intruders.

3.2 CONSTRUCTION OF NOVEL CHAOTIC SYSTEM

The introduced novel system is:

$$\dot{w}_1 = a(w_2 - w_1) + w_3 w_4$$
$$\dot{w}_2 = w_1(a - w_3) - w_2 + w_1 \quad (1)$$

$$\dot{w}_3 = w_1 w_2 - b w_3 + |w_3|$$
$$\dot{w}_4 = w_2 w_3 - w_4$$

Here, $w = (w_1, w_2, w_3, w_4)^T \in R^4$ are state variables and $a, b, c \in R$ are parameters. For parameter values $a = 10, b = 3, c = 20$ I.C. $(w_1(0), w_2(0), w_3(0), w_4(0)) = (0.5, 0.5, 0.5, 0.5)$ the times series and phase portraits of Eqn. (1) are displayed in Figures 3.1 and 3.2, respectively.

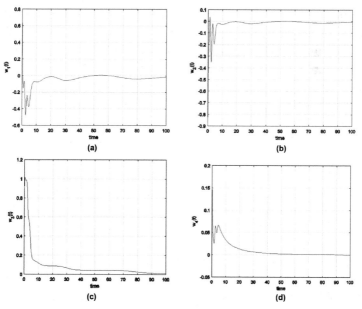

FIGURE 3.1 Time series of state variables w_1, w_2, w_3, w_4 of Eqn. (1), respectively.

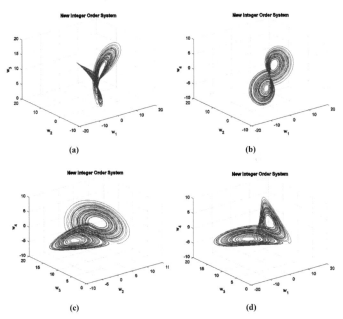

FIGURE 3.2 Phase portraits of Eqn. (1) in different planes.

3.3 QUALITATIVE ANALYSIS OF THE NEW CHAOTIC SYSTEM

We analyze the dynamics of the new chaotic system by plotting time series, phase plots, observing the existence and uniqueness of solution, checking the symmetric and dissipative behavior, undertaking the equilibrium point analysis, L.E. dynamics, bifurcation diagram, K.Y. (Kaplan-Yorke) dimension, etc.

3.3.1 ASYMMETRY

On changing the sign from to + to − and − to + of the state variables of the new system, we verify the asymmetric behavior the system possesses about all axis.

3.3.2 DISSIPATIVITIY

System (1) in matrix notation can be written as:

$$\begin{bmatrix} \dot{w}_1 \\ \dot{w}_2 \\ \dot{w}_3 \\ \dot{w}_4 \end{bmatrix} = \begin{bmatrix} F_1(w_1, w_2, w_3, w_4) \\ F_2(w_1, w_2, w_3, w_4) \\ F_3(w_1, w_2, w_3, w_4) \\ F_4(w_1, w_2, w_3, w_4) \end{bmatrix}$$

where;

$$F_1(w_1, w_2, w_3, w_4) = a(w_2 - w_1) + w_3 w_4$$
$$F_2(w_1, w_2, w_3, w_4) = w_1(a - w_3) - w_2 + w_1$$
$$F_3(w_1, w_2, w_3, w_4) = w_1 w_2 - bw_3 + |w_3|$$
$$F_4(w_1, w_2, w_3, w_4) = w_2 w_3 - cw_4$$

We check the dissipative character of the system by calculating ∇F as:

$$\nabla F = \frac{\partial F_1}{\partial w_1} + \frac{\partial F_2}{\partial w_2} + \frac{\partial F_3}{\partial w_3} + \frac{\partial F_4}{\partial w_4}$$
$$= -a - 1 - b + 1 - c$$
$$-33 < 0$$

Clearly, the described system is dissipative for any positive parameter values $a, b, c \in R$. For $a = 10$, $b = 3$, $c = 20$, using Liouville's theorem over a region $\Omega(t)$, we have:

$$\dot{V}(t) = \int_{\Omega(t)} (\nabla F) dw_1 dw_2 dw_3 dw_4 \tag{2}$$

where; volume is denoted by $V(t)$. Then:

$$\dot{V}(t) = \int_{\Omega(t)} (-33) dw_1 dw_2 dw_3 dw_4 = -33 V(t) \tag{3}$$

Integrating Eqn. (3), we get:

$$V(t) = V_0 e^{(-33t)}$$

for initial volume V_0, implying that with time the trajectories of Eqn. (1) evolve towards an attractor.

3.3.3 EXISTENCE AND UNIQUENESS OF SOLUTION

The novel 4-D chaotic system can be expressed as:

$$\dot{W}(t) = F(W(t))$$

where; $t \in (0, T)$ and the I.C. are given by $W(0) = W_0$.

Here, $W = \begin{bmatrix} w_1 \\ w_2 \\ w_3 \\ w_4 \end{bmatrix}$, $W_0 = \begin{bmatrix} w_{10} \\ w_{20} \\ w_{30} \\ w_{40} \end{bmatrix}$, $\Psi(Y(t)) = \begin{bmatrix} a(w_2 - w_1) + w_3 w_4 \\ w_1(a - w_3) - w_2 + w_1 \\ w_1 w_2 - bw_3 + |w_3| \\ w_2 w_3 - cw_4 \end{bmatrix}$

We now examine the solution for the system in the region $\Omega \times I$, where $I = (0, T]$, and $\Omega = (w_1, w_2, w_3, w_4)$: $\max |w_1|, |w_2|, |w_3|$ and $|w_4| \leq K, K > 0$. Parameter K lays a boundary for examining the existence and uniqueness of solution in the phase space.

The I.V.P. can be expressed as:

$$W(t) = W_0 + \int_0^t F(W(s))ds$$

Denote the quantity $W_0 + \int_0^t F(W(s))ds$ by $H(W)$, $W_1 = \begin{bmatrix} w_{11} \\ w_{12} \\ w_{13} \\ w_{14} \end{bmatrix}$, $W_2 = \begin{bmatrix} w_{21} \\ w_{22} \\ w_{23} \\ w_{24} \end{bmatrix}$, we get:

$$H(W_1) - H(W_2) = \int_0^t (F(W_1(s)) - F(W_2(s)))ds$$

This implies: $|H(W_1) - H(W_2)| = |\int_0^t (F(W_1(s)) - F(W_2(s)))ds|$

For $f(t) \in C(0, T]$, we consider here the norm $\|f\| = \sup_{t \in (0, T)} |f(t)|$.

For the matrix $P = |p_{ij}(t)|$ of continuous functions, we consider the norm $\|P\| = \sum_{i,j} \sup_{t \in (0,T]} |p_{i,j}(t)|$.

$$\|H(W_1) - H(W_2)\| \leq T \max(2|a|+1+2K, |a|+1+2K, |a|+1+2K, |b|+1+3K, |c|+K)\|W_1 - W_2\| \leq K_1 \|W_1 - W_2\| \quad (4)$$

where;

$$K_1 = T\max(2|a|+1+2K, |a|+1+2K, |b|+1+3K, |c|+K)$$

This W = H(W) is a contraction mapping for sufficient condition $0 < K_1 < 1$.

3.3.4 CONTINUOUS DEPENDENCE ON I.C.

We consider two I.C. W_{01} and W_{02} of system $\dot{W}(t) = F(W(t))$ such that:

$$\|W_0 - W_{02}\| \leq \delta$$

For the condition Eqn. (4), consider:

$$W_1 = W_{01} + \int_0^t F(W_1)(s)ds$$

$$W_2 = W_{02} + \int_0^t F(W_2)(s)ds$$

We get the following:

$$(1-K)\|W_1 - W_2\| \leq \|W_{01} - W_{02}\|$$

where; $0 < K < 1$. Let $\epsilon = \dfrac{\delta}{(1-K)}$, then:

$$\|W_1 - W_2\| \leq \epsilon$$

> **Theorem 3.1:** The solution for the novel system Eqn. (1) satisfying Eqn. (4) shows continuous dependence on I.C. if $\varepsilon > 0 \; \exists \; \delta(t) = (1-K_1)\varepsilon > 0$ such that $\|W_{01} - W_{02}\| \leq \delta$ implying $\|W_1 - W_2\| \leq \varepsilon$.

3.3.5 STABILITY OF EQUILIBRIUM POINTS

We determine the equilibrium points of Eqn. (1) by simply equating $F_i(w_1, w_2, w_3, w_4)$ to zero, i.e.:

$$a(w_2 - w_1) + w_3 w_4 = 0$$
$$w_1(a - w_3) - w_2 + w_1 = 0$$
$$w_1 w_2 - bw_3 + |w_3| = 0$$
$$w_2 w_3 - cw_4 = 0$$

For $a = 10$, $b = 3$, $c = 20$ we obtain three equilibrium points E_1, E_2, E_3 as $(-5.63743, -3.67144, 10.3487, -1.89974)$, $(0, 0, 0, 0)$ $(5.63743, 3.67144, 10.3487, 1.89974)$.

➤ **Theorem 3.2:** For chosen parameters, the equilibrium points E_1 is a stable hyperbolic and E_2, E_3 are unstable hyperbolic.

➤ **Proof 3.2:** For $E_1 = (-5.63743, -3.67144, 10.3487, -1.89974)$, we get the Jacobian matrix as:

$$J = \begin{bmatrix} -10 & 10 & -1.89974 & 10.3487 \\ 0.6513 & -1 & 5.63743 & 0 \\ -3.67144 & -5.63743 & -2 & 0 \\ 0 & 10.3487 & -3.67144 & -20 \end{bmatrix}$$

The eigenvalues of J are:

$\mu_1 = -20$
$\mu_2 = -12.3276$
$\mu_3 = -0.33601 + 6.4723i$
$\mu_4 = -0.33601 - 6.4723i$

Here, μ_1, μ_2 are negative eigenvalues μ_3, μ_4 are complex eigenvalues with negative real part, which implies E_1 is a stable hyperbolic equilibrium point since all eigenvalues have non-zero real part.

For $E_2 = (0, 0, 0, 0)$, we get the Jacobian matrix as:

$$J = \begin{bmatrix} -10 & 10 & 0 & 0 \\ 11 & -1 & 0 & 0 \\ 0 & 0 & -2 & 0 \\ 0 & 0 & 0 & -20 \end{bmatrix}$$

The given values of J are:

$\mu_2 = -2$
$\mu_2 = -2$

$$\mu_3 = \frac{1}{2}(-11 - \sqrt{521})$$

$$\mu_4 = \frac{1}{2}(-11 + \sqrt{521})$$

Here, μ_1, μ_2, μ_3 are negative eigenvalues μ_4 is positive eigenvalue which implies E_2 is an unstable hyperbolic equilibrium point since all eigenvalues have non-zero real part.

For $E_3 = (5.63743, 3.67144, 10.3487, -1.89974)$, we get the Jacobian matrix as:

$$J = \begin{bmatrix} -10 & 10 & 1.89974 & 10.3487 \\ 0.6513 & -1 & -5.63743 & 0 \\ 3.67144 & 5.63743 & -2 & 0 \\ 0 & 10.3487 & 3.67144 & -20 \end{bmatrix}$$

The eigenvalues of J are:

$\mu_1 = -16.7237 + 1.58174i$

$\mu_2 = -16.7237 - 1.58174i$

$\mu_3 = 0.223669 + 6.55912i$

$\mu_4 = 0.223669 + 6.55912i$

Here, μ_1, μ_2 are complex eigenvalues with negative real part, μ_3, μ_4 are complex eigenvalues with positive real part which implies E_3 is unstable hyperbolic equilibrium point since all eigenvalues have non-zero real part.

3.3.6 L.E. AND K.Y. DIMENSION

A confirmatory test to ascertain that the system under consideration is chaotic or not is to find the L.E. of the system. If we get a positive L.E. value, we can claim that the system is a chaotic system. L.E. give the rate of separation of trajectories evolving from close I.C.

For parameter values $a = 10$, $b = 3$, $c = 20$ and I.C. (.5.5.5.5), the system shows the following L.E. dynamics depicted in Figure 3.3.

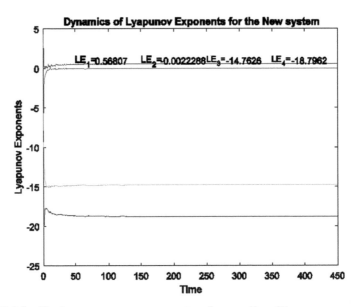

FIGURE 3.3 The Lyapunov exponent spectrum of system Eqn. (1).

The values of the L.E. as displayed in the figure are:

$$L_1 = 0.56807$$
$$L_2 = -0.0022288 \sim 0$$
$$L_3 = -14.7626$$
$$L_4 = -18.7962$$

Also the K.Y. dimension is:

$$D_{YK} = l + \frac{\sum_{i=1}^{l} L_i}{|L_{l+1}|}$$

where; l is the greatest number satisfying $\sum_{i=1}^{l} L_i \geq 0$ and $\sum_{i=1}^{l+1} L_i < 0$.

From the above values, the K.Y. dimension is 2.0168611, a non-integer value.

3.3.7 POINCARE SURFACE OF SECTION AND BIFURCATION ANALYSIS

The folding property of the chaotic system can be analyzed by plotting the Poincare surface of section we have plotted the Poincare sections in the x-y plane and z-w plane in Figure 3.4. The denseness of the points in these planes depict chaos in this system.

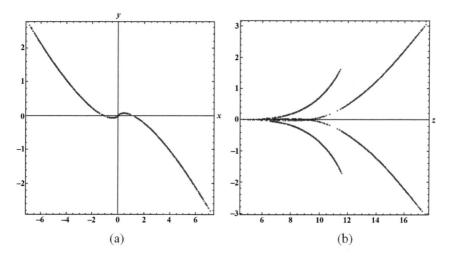

FIGURE 3.4 Poincare section of Eqn. (1) in different planes.

In bifurcation analysis, we vary a parameter keeping the other values constant. This helps observe the changing dynamics of the system.

Here for a = 10, b = 3, c = 20 and I.C. (0.5, 0.5, 0.5, 0.5) the bifurcation diagram are shown in Figure 3.5.

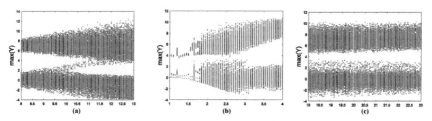

FIGURE 3.5 Bifurcation diagram of Eqn. (1) for: (a) $8 \leq a \leq 13$, b = 3, c = 20; (b) a= 1, $1 \leq b \leq 4$, c = 20; (c) a = 10, b = 3, $18 \leq c \leq 23$.

System's stability and equilibrium point analysis depend greatly on parametric values. Hence it is important to study the effect of change of parametric values on the existing states of the system.

Figure 3.5 represents the bifurcation diagram on changing the parameters. Figures 3.5(a) and 3.5(c) shows sustainable chaos for the range of parameter a and parameter c, respectively with only a slight variation. But Figure 3.5(b) shows undergoing bifurcations resulting in chaotic behavior of the system in the range $b \in (1, 4)$.

3.4 MULTI-SWITCHING PHASE SYNCHRONIZATION

To obtain the multi-switching phase synchronization between the identical novel 4-D double scroll system in the presence of uncertainties, we consider:

Master system:

$$\dot{w}_1 = a(w_2 - w_1) + w_3 w_4$$
$$\dot{w}_2 = w_1(a - w_3) - w_2 + w_1 \quad (5)$$

$$\dot{w}_3 = w_1 w_2 - bw_3 + |w_3|$$
$$\dot{w}_4 = w_2 w_3 - cw_4$$

where; $w = (w_1, w_2, w_3, w_4) \in R^4$ are the state variables of the system. For a = 10, b = 3, c = 20 and I.C. (0.5, 0.5, 0.5, 0.5), Eqn. (5) is chaotic.

Slave system:

$$\dot{z}_1 = a(z_2 - z_1) + z_3 z_4 + 0.5 z_4 + \psi_1$$
$$\dot{z}_2 = z_1(a - z_3) - z_2 + z_1 + \psi_2 \quad (6)$$

$$\dot{z}_3 = z_1 z_2 - bz_3 + |z_2| + \psi_3$$
$$\dot{z}_4 = z_2 z_3 - cz_4 + \psi_4$$

where; $z = (z_1, z_2, z_3, z_4) \in R^4$ are the state variables of the system. For a = 10, b = 3, c = 20 and I.C. (0.7. 08. 0.9, 0.7), Eqn. (6) is chaotic. Here

$\psi_i : i = 1, 2, 3, 4$ are controllers which will be designed to achieve synchronization. Figure 3.6 displays the phase portraits of Eqn. (6).

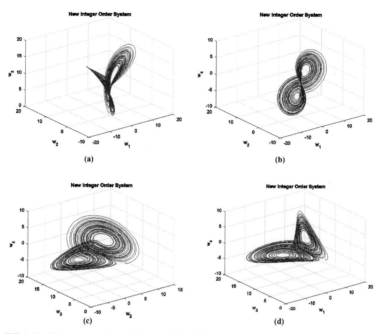

FIGURE 3.6 Phase portraits of Eqn. (6) in different planes.

3.4.1 DISLOCATED PHASE SYNCHRONIZATION

The multi-switching synchronization increases the flexibility of the synchronization scheme as the choice of existence of switches multiplies. Here randomly a switch is chosen as:

Switch:

$$\begin{aligned} e_1 &= z_1 - w_4 \\ e_2 &= z_2 - w_3 \\ e_3 &= z_3 - w_2 \\ e_4 &= z_4 - w_1 \end{aligned} \quad (7)$$

Differentiating Eqn. (7), we get:

$$\dot{e}_1 = \dot{z}_1 - \dot{w}_4$$
$$\dot{e}_2 = \dot{z}_2 - \dot{w}_3 \tag{8}$$

$$\dot{e}_3 = \dot{z}_3 - \dot{w}_2$$
$$\dot{e}_4 = \dot{z}_4 - \dot{w}_1$$

Using Eqns. (5) and (6) we have:

$$\dot{e}_1 = (a(z_2 - z_1) + z_3 z_4 + 0.5 z_4 + \psi_1) - (w_2 w_3 - c w_4)$$
$$\dot{e}_2 = (z_1(a - z_3) - z_2 + z_1 + \psi_2) - (w_1 w_2 - b w_3 + |w_3|) \tag{9}$$

$$\dot{e}_3 = (z_1 z_2 - b z_3 + |z_3| + \psi_3) - (w_1(a - w_3) - w_2 + w_1)$$
$$\dot{e}_4 = (z_2 z_3 - c z_4 + \psi_4) - (a(w_2 - w_1) + w_3 w_4)$$

Choosing the nonlinear control functions as:

$$\psi_1 = -10 z_2 + 10 z_1 - z_3 z_4 - 0.5 z_4 + w_2 w_3 - 20 w_4 - e_1$$
$$\psi_2 = -10 z_1 + z_1 z_3 + z_2 - z_1 - 0.3 z_3 + w_1 w_2 - 3 w_3 + |w_3| - 2 e_2$$
$$\psi_3 = -z_1 z_2 + 3 z_3 - |z_3| - 0.4 z_2 + 10 w_1 - w_1 w_3 - w_2 + w_1 - 3 e_3 \tag{10}$$
$$\psi_4 = -z_2 z_3 + 20 z_4 - 0.2 z_1 + 10 w_2 - 10 w_1 + w_3 w_4 - 4 e_4$$

Substituting Eqn. (10) into Eqn. (9), we get:

$$\dot{e}_1 = -e_1$$
$$\dot{e}_2 = -2 e_2$$
$$\dot{e}_3 = -3 e_3$$
$$\dot{e}_4 = -4 e_4$$

Next consider the positive definite Lyapunov function as $V(e(t)) = \frac{1}{2} e(t) e(t)^T$

$$= \frac{1}{2}(e_1^2 + e_2^2 + e_3^2 + e_4^2)$$

Secure Communication Using a Novel 4-D Double Scroll Chaotic System

$$\dot{V}(e(t)) \le e_1\dot{e}_1 + e_2\dot{e}_2 + e_3\dot{e}_3 + e_4\dot{e}_4$$
$$= e_1(-e_1) + e_2(-2e_2) + e_3(-3e_3) + e_4(-4e_4)$$
$$= -e_1^2 - 2e_2^2 - 3e_3^2 - 4e_4^2$$

where; $\dot{V}(e(t))$ is negative definite.

From Lyapunov stability theory we have $e_i \to 0$ for i = 1, 2, 3, as t ≥ ∞. Hence the trajectories of Eqns. (5) and (6) are synchronized in multi-switching phased manner as displayed in Figure 3.7.

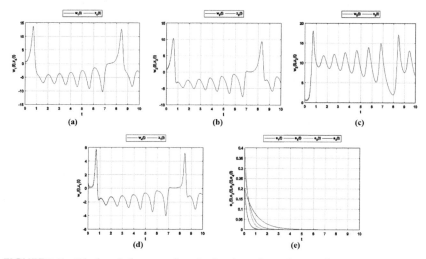

FIGURE 3.7 Disclosed phase synchronized trajectories and error plot.

3.5 APPLICATION IN SECURE COMMUNICATION

Chaotic systems are finding growing application in the field of secure communication, control system, image encryption, etc., owing to their entirely new dynamics even with the slightest change in parameter values and I.C.

This also leads to the development of new synchronization techniques to increase the robustness in its application.

Here we have utilized the novel synchronization technique viz. 'Multi-switching phase synchronization' on the novel system insecure communication. The idea is to mix the chaotic signals with the original signal and transmit. The original signal is then recovered after performing the required synchronization.

> **Illustration:** Let the original signal be $q(t) = \sin(t) - \cos(t)$. We chaotify it using $w_1(t)$ only to recover the signal later after performing synchronization, using the above-designed controller ψ_1. The results are displayed in Figure 3.8.

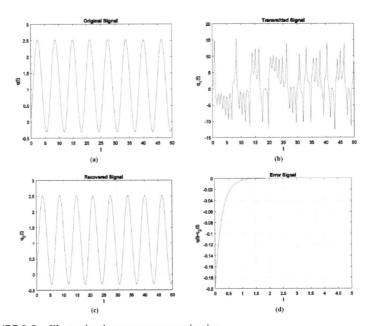

FIGURE 3.8 Illustration in secure communication.

3.6 CONCLUSION

In this chapter, a novel 4-D double scroll chaotic system with two unstable and one stable equilibrium point has been presented. The dynamical properties such as LEs, existence, and uniqueness of solution, dissipative, and symmetric character, its equilibrium points analysis has been studied in detail. Synchronization of the novel system in multi-switching phase synchronization is performed considering uncertainties. The application of the attained synchronization has been displayed in secure communication with the help of an illustration. These systems being non-standard systems are difficult to be recognized by the hackers and therefore would find suitable application across various disciplines such as secure communication, control systems, image encryption, cryptography, etc.

The future scope of these studies comprises of designing the circuit realization of the new introduced 4-D chaotic system and investigate its hidden attractors.

KEYWORDS

- dislocated phase synchronization
- dynamical analysis
- Lyapunov exponent
- nonlinear control functions
- novel chaotic system
- secure communication

REFERENCES

1. Khan, A., & Trikha, P., (2019). Compound difference anti-synchronization between chaotic systems of integer and fractional order. *SN Appl. Sci., 1*, 757. https://doi.org/10.1007/s42452-019-0776-x.
2. Wu, Z., Zhang, X., & Zhong, X., (2019). Generalized chaos synchronization circuit simulation and asymmetric image encryption. *IEEE Access*.
3. Khan, A., & Trikha, P., (2020). Study of earth's changing polarity using compound difference synchronization. *GEM-International Journal on Geomathematics, 11*(1), 7. Springer.
4. Khan, A., Jahanzaib, L. S., Nasreen, T. P., & Khan, T., (2020). Compound difference anti- synchronization between hyper-chaotic systems of fractional order. *Journal of Scientific Research, 12*(2), 175–181.
5. Trikha, P. N., & Jahanzaib, L. S., (2020). Combination difference synchronization between identical generalized Lotka-Volterra chaotic systems. *Journal of Scientific Research, 11*(2), 183–188.
6. Khan, A., Trikha, P., & Jahanzaib, L. S., (2019). Double compound combination anti-synchronization in a non-identical fractional-order hyperchaotic system. *Journal of Basic and Applied Engineering Research, 6*(8), 431–436.
7. Khan, A., Jahanzaib, L. S., & Trikha, P., (2019). Dual combination anti-synchronization of non-identical fractional-order chaotic system with different dimension using scaling matrix. *Journal of Basic and Applied Engineering Research, 6*(8), 437–443.
8. Li, C., & Sprott, J. C., (2014). Multistability in the Lorenz system: A broken butterfly. *International Journal of Bifurcation and Chaos, 24*(10), 131, 1450.

9. Li, C., Sprott, J. C., & Thio, W., (2015). Linearization of the Lorenz system. *Physics Letters A, 379*(10, 11), 888–893.
10. Ahmadi, A., Rajagopal, K., Alsaadi, F. E., Viet-Thanh, P., Alsaadi, F. E., & Jafari, S. A., (2019). Novel 5D chaotic system with extreme multi-stability and a line of equilibrium and its engineering applications: Circuit design and FPGA implementation. *Iranian Journal of Science and Technology, Transactions of Electrical Engineering*. Springer.
11. Jafari, S., Pham, V. T., & Kapitaniak, T., (2016). Multi scroll chaotic sea obtained from a simple 3D system without equilibrium. *International Journal of Bifurcation and Chaos, 26*(02), 031, 1650.
12. Jafari, S., Sprott, J. C., & Molaie, M., (2016). A simple chaotic ow with a plane of equilibria. *International Journal of Bifurcation and Chaos, 26*(06), 098, 1650.
13. Li, C., Hu, W., Sprott, J. C., & Wang, X., (2015). Multistability in symmetric chaotic systems. *The European Physical Journal Special Topics, 224*(8), 1493–1506.
14. Rajagopal, K., Durdu, A., Jafari, S., Uyaroglu, Y., Karthikeyan, A., & Akgul, A., (2019). Multiscroll chaotic system with sigmoid nonlinearity and its fractional-order form with synchronization application. *International Journal of Non-Linear Mechanics, 116*. Elsevier.
15. Singh, J. P., Rajagopal, K., & Roy, B. K., (2018). A new 5D hyperchaotic system with stable equilibrium point, transient chaotic behavior, and its fractional-order form. *Pramana*. Springer.
16. Vaidyanathan, S., Volos, C. K., Kyprianidis, I., Stouboulos, I., & Pham, V. T., (2015). Analysis, adaptive control, and anti-synchronization of a six-term novel jerk chaotic system with two exponential nonlinearities and its circuit simulation. *Journal of Engineering Science and Technology Review, 8*(2).
17. Pham, V. T., Jafari, S., Wang, X., & Ma, J., (2016). A chaotic system with different shapes of equilibria. *International Journal of Bifurcation and Chaos, 26*(04), 069, 1650.
18. Viet-Thanh, P., Vaidyanathan, S., Volos, C., & Kapitaniak, T., (2018). *Nonlinear Dynamical Systems with Self-Excited and Hidden Attractors, 133*. Springer.
19. Wu, F., Zhang, G., & Ma Jun J. A., (2019). A neural memristor system with infinite or without equilibrium. *The European Physical Journal Special Topics, 228*. Springer.
20. Wolf, A., Swift, J. B., Swinney, H. L., & Vastano, J. A., (1985). Determining Lyapunov exponents from a time series. *Physica D: Nonlinear Phenomena, 16*(3), 285–317.
21. Khan, A., Khattar, D., & Prajapati, N., (2017). Adaptive multi switching combination synchronization of chaotic systems with unknown parameters. *International Journal of Dynamics and Control*, 1–9.
22. Khan, A., & Tyagi, A., (2018). Disturbance observer-based adaptive sliding mode hybrid projective synchronization of identical fractional-order financial systems. *Pramana, 90*(5), 67.
23. Khan, A., et al., (2017). Increased and reduced-order synchronizations between 5D and 6D hyperchaotic systems. *Indian Journal of Industrial and Applied Mathematics, 8*(1), 118–131.
24. Dudkowski, D., Jafari, S., Kapitaniak, T., Kuznetsov, N. V., Leonov, G. A., & Prasad, A., (2016). Hidden attractors in dynamical systems. *Physics Reports, 637*. Elsevier.
25. Rajagopal, K., Akgul, A., Moroz, I. M., Wei, Z., Jafari, S., & Hussain, I., (2019). A simple chaotic system with topologically different attractors. *IEEE Access, 7*.

CHAPTER 4

DETECTING HATE SPEECH THROUGH MACHINE LEARNING

F. H. A. SHIBLY,[1] UZZAL SHARMA,[2] and H. M. M. NALEER[3]

[1]*Assam Don Bosco University/South Eastern University of Sri Lanka, Sri Lanka, E-mail: shiblyfh@seu.ac.lk*

[2]*Assam Don Bosco University, Assam, India*

[3]*South Eastern University of Sri Lanka, Sri Lanka*

ABSTRACT

Hatred and abusive speeches are identified as huge crime that have been incrementing very recent years, and this has been not only in the specific interaction done face to face but preferably also in the online sharing of information. A considerable number of factors have been contributing to this. There has been a specific study that has well provided a kind of critical overview on the way of detecting such speeches within post or text has been highly evolved over the last few years. However, it has been observed that there are few studies that have been published in detecting hate speech automatically from the perspective of computer science. There is a great way or process with the support of which hate speech can be detected. It has been observed that both the automatic speech recognition and machine learning (ML) have been together complementing each other in the current past, and this has been because both the paradigms are very much deeply ingrained in each other. Therefore, this chapter aims to find out the relationship between hate speech detection and ML and find out the feasible ML algorithms to control hate speeches in social media. It has been observed that there is the implementation of a huge range of several methods of classifying the utilization of the embedding learning for computing all the distances which are semantic in between various parts of the speech which is properly considered to be a specific part of the "othering" narrative. It has also been observed that both the automatic recognition of speech and the ML have been hugely complementing each other in the current past, and this has been just because of the fact that both the paradigms are very much deeply ingrained in each other.

4.1 INTRODUCTION

Hate crimes are becoming a significant issue in social media nowadays. Apart from social media, it has expanded in people's day-to-day life. There are many incidents have been recorded related to the hate crimes all over the world. The open space of social media and the collaborative opportunity offers a significant opportunity for exchanging of abusive speeches against a person or group or community or anyone. These opinions are stated not only in social media but also in electronic media, Online forums, discussion groups as well as in blog websites. There has been a specific study that has well provided a kind of critical overview on the way an automatic detection of hate speech within the text has highly evolved over the last few years. However, it has been observed that there are few studies that have been published in the automatic detection of hate speech from the perspective of computer science. There is a great way or process with the help of which hate speech can be detected. A new method has been proposed for identifying hate speech on Twitter [9].

Hate speech is a huge crime that has been incrementing in the current years, and this has been not only in the specific interaction done face to face but preferably also in the online sharing of information [5]. A considerable number of factors have been contributing to this. There has been a specific study that has well provided a kind of critical overview on the way an automatic detection of hate speech within the text has been highly evolved over the last few years. However, it has been observed that there are few studies that have been published in the automatic detection of hate speech from the perspective of computer science. There is a great way or process with the help of which hate speech can be detected. A new method has been proposed for detecting hate speech on Twitter [9]. This approach will be automatically detecting all the various patterns of hate speech and the most common unigrams and utilize all of these together will all the various semantic as well as the sentimental features for classifying various tweets into offensive and also clean.

Offensive or rather any antagonistic language is targeted at all the various individuals as well as the social groups which will be based upon all of their very personal characteristics and it has been very much frequently posed as well as broadly spread via the World Wide Web. This can be well-thought-out to be the main key factor for the individual and also the tension of the society surrounding the instability of the region. Hence, it has been thought that automated cyberhate detection based upon the web is very much essential for both understandings as well as observing all the societal tension of the regions, especially in all the online social networks where several posts can be hugely disseminated as well as viewed. While several previous works have been involved utilizing the bag of words or rather lexicons, they will be suffering from the same kind of matter, which is actually that cyberhate can be both incidental as well as subtle. Thus, depending upon the particular occurrence of all the various arguments or rather sayings can be leading to a weighty number of several false denials, giving an improper depiction of the trends within the cyberhate [1]. This particular problem has seen to be motivating the specific challenge thinking regarding the specific representation of the use of the subtle language like the references to all the several perceived threats from "the other" set of feature that will be utilizing the use of language around the particular concept of "othering" and the theory of the intergroup threat for identifying all the various subtleties. There is the implementation of a broad range of several methods of classification utilizing the embedding learning for computing all the semantic spaces among various parts of the speech properly considered to be a part of the "othering" narrative. For validating such a kind of approach, there are two experiments that have been conducted. The first one will be involving the comparison of all the outcomes of the novel method with the specific state of the models of the art's baseline. The second one has actually tested all the models which have been best performing from the very first stage on all the datasets which have been unseen for several kinds of cyberhate, namely, orientation, race, and disability. There are a number of proposals, but among others, there is a great proposal of a novel 'othering' set of feature that will be actually using the language utilization around the specific concept of 'othering' and also an intergroup theory of threat for properly identifying all of such subtleties. There is also a great implementation of a broad range of several methods of classification utilizing the learning which has been embedded for directly computing several semantic distances in between several parts of speech which have been individually considered to be a particular part of the narrative of 'othering.'

4.2 DETECTION OF HATE SPEECH THROUGH MACHINE LEARNING (ML)

It has been observed that both automatic speech recognition and machine learning (ML) have been together complementing each other in the current past, and this has been because both the paradigms are very much deeply ingrained in each other. ML in the ASR can be initiated with that of the ANN-based recognizer of speech and then finally followed by several hybrid systems of either HMM or ANN. However, there may be a loss of momentum because of the huge difficulty within all the adopted techniques of learning [7]. However, it is to be well known that designing an efficient architecture of deep learning as well as algorithms that will be scalable as well as robust with incomplete data is a significant challenge. It will be very much important to compare the speech of hate with that of cyberbullying, flaming, radicalization, and a language which will be abusive. As this kind of prejudiced sharing of information can be very much harmful to the society, all the platforms of social networks and the governments can greatly benefit from both the prevention as well as the detection [6]. It is to be well known that both the systematization and the development of all the various resources which have been shared, like the guidelines as well as algorithms are really a very much crucial step in the advancement of automatic hate speech detection.

As it has been known that it is really an essential problem of discerning the content which will be hateful within the social media, a detection scheme has been observed to be proposed, which is basically the ensemble of the recurrent neural network (RNN) classifiers. It will be including several features that will be directly associated with the information which will be user-related, like the tendency of the users towards either sexism or rather a racism. Such kinds of data are fed as input to all the classifiers together with the vectors of the word frequency, which have been derived from that of the text content. Automated detection of the language which is very much abusive in the online media, has in recent years, become the main challenge. In one paper, there has been a well presentation of an ensemble classifier for detecting the hate-speech in very much short test like that of the tweets [2]. The classifier utilized deep learning and includes a specific series of several features that have been associated with several behavioral characteristics of the users, like the tendency of posting several abusive messages as the input to that of the classifier. There are a number of various contributions for particularly advancing the art's state. Firstly, there is the development of an architecture of deep learning that will be utilizing the vectorization of the frequency of the word for the implementation of all the various features. Secondly, there is also a method that has been proposed independently of any language because of the no-usage of all the embedding of the pre-trained words. Thirdly, there has been a proper thorough evaluation of the model utilizing a set of the public data of several labeled tweets, an implementation that is fully open-sourced built upon the peak of the Keras. This particular estimation will also be involving a specific analysis of the proposed scheme's performance for the different classes of the users. It has been observed that there are some possible threats as well to that of the vulnerability and various limitations of the approach. The kind of behavior which is stochastic of the processes of deep learning is considered to be the most essential threat for constructing the validity, and this has resulted in the fluctuation in the particular F-score over all the various runs. For particularly overcoming this, there are various experiments as well which have been observed. One approach has been made on a corpus of 16k tweets, which has been made public and ultimate results will be demonstrating the effectiveness as compared to all the existing solutions of state of the art. For specifically solving the problem of the detection of hate speech, there is also a learning of the distributed comment's representation which will be low dimensional utilizing the currently proposed models of neural language which can be directly fed as particular inputs to an algorithm of classification [4]. There is a two-step method of the detection of speech. Firstly, there is the utilization of paragraph2vec for the specific joint modeling of both words as well as comments where everyone will be able to learn all of their distributed representations in the specific joint space utilizing the continuous neural language model of BOW. This will be actually resulting in the text-embedding which will be very much low-dimensional where similar kinds of comments, as well as words, can be residing within the same specific part of the space. Secondly, there is the utilization of embedding for training a specific binary classifier for particularly differentiating the clean and the hateful comments. During the inference, for all the newly observed comments, we will be actually inferring the specific representation by directly "folding in" utilizing all the embedding that have been learned already. In detail, it will be involving basically two parts-neural language model and the empirical analysis. Models of neural language will be taking full advantage of the order of the word and stating the similar kind of assumption of the n-gram model of language that all the variables which are very much closer within a sentence are much more statistically dependent. The CBOW model has been actually utilized as a specific kind of component of the paragraph2vec based upon the words which have been surrounding actually tries a lot to predict the central word directly and also the comment of the user the words belong to. As per the empirical analysis will be considered, the set of data will be comprised of about 56,280 comments involving the hate speech and about 895.456 full clean comments which have been generated by about 209,776 anonymized users which have been collected over a period of 6 months. There is also a full new theoretical structure for underpinning the algorithms of the detection of hate speech on the internet [3]. The principles of crime pattern theory and all the conceptualizations of the cyber place dependent upon the specific digital spaces of the particular convergence have been well adapted for particular identifying all the various characteristics which have been well associated with the dissemination of hate speech in Twitter. There are basically two relevant cyber places that have been well-identified for the dissemination of the hate speech on Twitter, which involves the accounts and the tweets. Specifically drawing on the particular technique of the Random Forests, the tweet Metadata have been proved to be very much efficient in the hate speech classification.

It has also been observed that all the offensive, as well as the abusive languages, have been well identified in Indonesian Twitter utilizing the approach of deep learning. There is the utilization of an LSTM or long short-term memory with the embedding of a word as it has been found that LSTM with the embedding of expression is really very much useful for the classification of text. Hence, it has been made very much clear from here that there are various current advances in the technology, particularly deep learning, which is an area of ML capable of finding a number of applications in the analytics of big data and AI. Digital metadata of the microenvironment can be utilized for detecting several patterns of hate speech in the cyberspace, which is very much similar to that of the specific way all the models of spatiotemporal crime can be found in the physical environment [8]. On the other hand, messages of hate speech on Twitter will be describing several environmental patterns that will be very much different from all the messages which are neutral. This particular result has been actually derived from the basic fact that several messages of hate speech are well communicated via all the tweets or rather through various accounts with particular environmental features reflected in the very much concrete metadata which will be associated with that of the message. Table 4.1 provides a good description regarding the contribution of ML in the detection of hate speech.

TABLE 4.1 Features and Challenges of Hate Speech Detection

Authors	Research Method	Features	Issues and Challenges
Mossie et al.	GRU and RNN	• Eliminate subjects that lead emotional issues • High AUC and accuracy	• Need to handle negations • Can happen on multiple issues and different social networking sites
Al-Makhadmeh and Amr	KNLPEDNN	• Achieves minimum loss function values • Minimizes the deviations	• Need to improve the ensemble learner process
Pitsilis et al.	RNN	• Achieve higher classification quality	• Dataset is imbalanced • Fluctuation in the f-score
Şahi et al.	SVM	• Improves recall • Increases the precision	• Misclassification needs to be diminished
Pratiwi et al.	FastText classifier	• Reduce variations of incomplete words	• Lowest f-measure • Not suitable for small dataset • Unable to work optimally

TABLE 4.1 *(Continued)*

Authors	Research Method	Features	Issues and Challenges
Zhang et al.	Deep Learning	• Reduce errors proportionally	• Relatively small improvement
			• Significantly higher errors
Rohmawati et al.	SEMAR	• Highest average accuracy score	• Category is misclassified
Albadi et al.	GRU neural networks with handcrafted features	• Best recall • Best at minimizing false negatives	• Requires enhancement in classification accuracy.

4.3 CONCLUSION

We finally found that detection of hate speech is possible by utilizing various kinds of ML techniques. It has been observed that there is the implementation of a huge range of several methods of classifying the utilization of the embedding learning for computing all the distances which are semantic in between various parts of the speech which is properly considered to be a specific part of the "othering" narrative. It has also been observed that both the automatic recognition of speech and ML have been hugely complementing each other in the current past, and this has been just because of the fact that both the paradigms are very much deeply ingrained in each other. The ensemble of the classifiers of RNN and the model of CBOW are some of the evidence supporting the fact that detection of hate speech is possible utilizing all the several techniques of ML even though it is considered to be very much challenging. The particular utilization of an LSTM with that of the embedding of the word says about the contribution of it in the detection. It has been found that LSTM with the embedding of the word is really very much useful for the classification of text. Such concepts from the literature review provide with a lot of information regarding the fact that several ML techniques are utilized for the detection of hate speech.

KEYWORDS

- hate speech
- long short-term memory
- machine learning
- recurrent neural network
- social media
- worldwide web

REFERENCES

1. Alorainy, W., Burnap, P., Liu, H., & Williams, M. L., (2019). The enemy among us: Detecting cyber hate speech with threats-based othering language embeddings. *ACM Transactions on the Web (TWEB)*, *13*(3), 14.
2. Djuric, N., Zhou, J., Morris, R., Grbovic, M., Radosavljevic, V., & Bhamidipati, N., (2015). Hate speech detection with comment embeddings. In: *Proceedings of the 24th International Conference on World Wide Web* (pp. 29, 30). ACM.
3. Ibrohim, M. O., Sazany, E., & Budi, I., (2019). Identify abusive and offensive language in Indonesian twitter using deep learning approach. In: *Journal of Physics: Conference Series* (Vol. 1196, No. 1, p. 012041). IOP Publishing.
4. Miró-Llinares, F., Moneva, A., & Esteve, M., (2018). Hate is in the air! But where? Introducing an algorithm to detect hate speech in digital microenvironments. *Crime Science*, *7*(1), 15.
5. Fortuna, P., & Nunes, S., (2018). A Survey on Automatic Detection of Hate Speech in Text. *ACM Computing Surveys*, *51*(4), 1–30.
6. Padmanabhan, J., & Johnson, P. M. J., (2015). Machine learning in automatic speech recognition: A survey. *IETE Technical Review*, *32*(4), 240–251.
7. Pitsilis, G. K., Ramampiaro, H., & Langseth, H., (2018). Effective hate-speech detection in Twitter data using recurrent neural networks. *Applied Intelligence*, *48*(12), 4730–4742.
8. Simon, A., Mahima, S. D., Venkatesan, S., & Babu, R. D. R., (2016). An overview of machine learning and its applications. *Int. J. Electr. Sci. Eng.*, *1*, 22–24.
9. Watanabe, H., Bouazizi, M., & Ohtsuki, T., (2018). Hate speech on Twitter: A pragmatic approach to collect hateful and offensive expressions and perform hate speech detection. *IEEE Access*, *6*, 13825–13835.
10. Pratiwi, N. I., Budi, I., & Alfina, I., (2018). Hate speech detection on Indonesian Instagram comments using fast text approach. In: *2018 International Conference on Advanced Computer Science and Information Systems (ICACSIS)* (pp. 447–450). Yogyakarta.
11. Ziqi Z., David, R., & Jonathan, T., (2018). *Detecting Hate Speech on Twitter Using A Convolution-GRU Based Deep Neural Network* (pp. 745–760). The Semantic Web.

12. Rohmawati, U. A. N., Sihwi, S. W., & Cahyani, D. E., (2018). SEMAR: An interface for Indonesian hate speech detection using machine learning. In: *2018 International Seminar on Research of Information Technology and Intelligent Systems (ISRITI)* (pp. 646–651). Yogyakarta, Indonesia.
13. Albadi, N., Kurdi, M., & Mishra, S., (2019). Investigating the effect of combining GRU neural networks with handcrafted features for religious hatred detection on Arabic Twitter space. *Social Network Analysis and Mining, 9*(1).
14. Zewdie, M., & Jenq-Haur, W., (2019). Vulnerable community identification using hate speech detection on social media. *Information Processing and Management, in Communication.*
15. Al-Makhadmeh, Z., & Tolba, A., (2019). Automatic hate speech detection using killer natural language processing optimizing ensemble deep learning approach. *Computing, 102*(2), 501–522.
16. Georgios, K. P., Heri, R., & Helge, L., (2018). Effective hate-speech detection in Twitter data using recurrent neural networks. *Applied Intelligence.*
17. Şahi, H., Kiliç, Y., & Sağlam, R. B., (2018). Automated detection of hate speech towards woman on Twitter. In: *2018 3rd International Conference on Computer Science and Engineering (UBMK)* (pp. 533–536). Sarajevo.

CHAPTER 5

OPTIMIZATION OF LOGICAL RESOURCES IN RECONFIGURABLE COMPUTING

S. JAMUNA

Department of Electronics and Communication Engineering, Dayananda Sagar College of Engineering, VTU, Bangalore, Karnataka, India, E-mail: jamuna-ece@dayanandasagar.edu

ABSTRACT

A computing system which can be reconfigured depending on the functionality requirement is called as a reconfigurable computing system. Such kind of a computing system can be easily implemented using reconfigurable device, i.e., a reconfigurable field programmable gate arrays (FPGAs). Computing systems are designed using reconfigurable devices so that they can adapt and evolve in response to the changing application environment while executing the assigned task. At present FPGAs are used in almost all applications like space missions, communication systems, automotive and industrial applications etc. In any digital system design, it is very important to optimize parameters such as processing time, power consumption, memory requirements and hardware resources, etc. In FPGAs number of logic resources utilized will depend upon the complexity of the application design. Since the availability of logic resources is fixed for any FPGA, there is an option for the designer to utilize efficiently. Run time re-configurability feature available in recent FPGAs helps the designer to implement multiple functionalities with time multiplexed hardware. Partial reconfiguration facilitates run-time re-configurability. SoC functional blocks mapped on high-end FPGAs can be partially reconfigured without affecting system's normal functionality. PR design methodology enables new types of FPGA designs that provide efficiencies unattainable with conventional design techniques [1]. In this chapter an attempt is made to explain how the number of logical resources utilization can be reduced through partial reconfiguration. Three case study applications are illustrated in support of this. Design and implementation of ALU block, the modulation functional blocks and encryption algorithms on FPGA are explained. Design is been done using XILINX Vivado software and implemented on ZED board.

5.1 INTRODUCTION

A computing system that can be reconfigured depending on the functionality requirement is called a reconfigurable computing system. Such kind of a computing system can be easily implemented using the reconfigurable device, i.e., a reconfigurable field-programmable gate arrays (FPGAs). Computing systems are designed using reconfigurable devices so that they can adapt and evolve in response to the changing application environment while executing the assigned task. At present, FPGAs are used in almost all applications like space missions, communication systems, automotive, and industrial applications, etc. In any digital system design, it is very important to optimize parameters such as processing time, power consumption, memory requirements and hardware resources, etc. In FPGAs number of logic resources utilized will depend upon the complexity of the application design. Since the availability of logic resources is fixed for any FPGA, there is an option for the designer to utilize efficiently. Run time re-configurability feature available in recent FPGAs helps the designer to implement multiple functionalities with time-multiplexed hardware. Partial reconfiguration (PR) facilitates runtime re-configurability. Functional blocks of SoCs mapped on high-end FPGAs can be partially reconfigured without affecting the system's normal response. PR design methodology enables new types of FPGA designs that provide efficiencies unattainable with conventional design techniques [1]. In this chapter, an attempt is made to explain how the number of logical resources utilization can be reduced through PR. Three case study applications are illustrated in support of this. Design and implementation of ALU block, the modulation functional blocks and encryption algorithms on FPGA are explained. XILINX Vivado software is used for design and is implemented on ZedBoard.

FPGAs are the flexible hardware structures used for implementing a wide variety of applications starting from basic logic functions to complex system-on-chips (SOC). The device gets programmed depending on the design and the bit file generated during the implementation process. The concept of programmability varies from device to device. In SRAM-based FPGAs, required functionality will be implemented by downloading bit file on to the configuration memory. The flexibility of FPGAs further can be extended with the help of PR, i.e., by modifying FPGA designs partially. PR allows a particular selected region of FPGA to dynamically reconfigure, without altering the functionality of other designs on FPGA. This is achieved with the help of partial bit file. PR can be used even for complex FPGA design implementation. The advantages of the PR design methodology are flexibility, efficient usage of device area, and less reconfiguration time because of smaller bit files.

This chapter illustrates an implementation of three important applications based on partial reconfigurable design methodology. In all the three cases, logical resources have been reduced, and comparison is given. The first functional block is arithmetic and logic unit (ALU), an essential sub-block found in most of the digital computing systems, especially microprocessors/controllers, since they involve in computing activities. The first and foremost important task of an ALU is to perform arithmetic and logical operations [2]. Microprocessors include standalone design to perform arithmetic operations and its speed also mainly depends on ALU performance. In complex data computation systems, ALU is implemented with various bit-widths to increase its efficiency [3, 4]. Since the size of ALU is decreasing and its complexity is increasing, a sophisticated method should be acquired while designing ALU, such that resources utilization and power consumption can be optimized. PR helps in optimizing ALU design.

The second case study application is the modulator block used in communication systems. In recent years, wireless communication is dominantly preferred over wired communication because of its advantages like *ease of use, flexibility, cost of maintenance, durability*, etc. Different modulation techniques used are binary phase-shift keying (BPSK), binary frequency shift keying (BFSK), binary amplitude shift keying (BASK), quadrature amplitude modulation (QAM), and quadrature phase-shift keying (QPSK) which modulate the source data that is to be transmitted [5]. Instead of design and implementing individual modulators, user can dynamically opt between different modulation techniques with respect to the requirements of the communication block. So here also, PR can be used for changing different models dynamically within the active design. This design mainly deals with BPSK, QPSK, and 16-QAM modulation techniques which are implemented with the help of PR to achieve optimization in resource utilization and power consumption.

The third application considered is the data encryption algorithm. In data communication, safety, and confidentiality of data is achieved through a suitable encryption algorithm. Encryption is the most important aspect when it comes to security [6]. There are various types of encryption algorithms to communicate data securely. Design flexibility can be increased by providing an option to the user to select a particular algorithm as per the requirement. Instead of using a single algorithm to encrypt data, multiple algorithms can be used with an option to switch between the algorithms. But the resources used will be the same. Depending on the selection, time-multiplexed reconfiguration will be done. Thus optimizing the resource utilization and also can avoid security breach. In this chapter two encryption algorithms, i.e., advance encryption standard (AES) and TwoFish both of 128-bit are chosen to reconfigure at runtime using a custom ICAP (internal configuration access port) controller IP provided by Xilinx and is implemented on ZedBoard. The main advantage of this implementation is that the user have an option to switch between two algorithms.

5.2 BACKGROUND DETAILS

Reconfigurable computing systems use FPGAs for hardware implementation. An FPGA is a regular 2D structure of logic resources (CLB) and interconnections, which can be controlled by the user control. These can be programmed accordingly as per requirements to achieve desired circuit. Unlike the old generation FPGAs, today's FPGAs are fabricated with different configurable embedded SRAM, high-speed I/Os and transceivers functional blocks. These FPGAs are also including logic blocks that include memory elements varying from simple flip flop to more complex memory blocks.

5.2.1 OVERVIEW OF FPGA ARCHITECTURE

Figure 5.1 shows the detailed overview of FPGA architecture. In general, the SRAM-based FPGA architecture includes N rows and N columns of configurable logic blocks (CLBs) [7]. These CLBs can be programmed with configuration data that implements any logic functions. A programmable interconnect network connects all the CLBs individually.

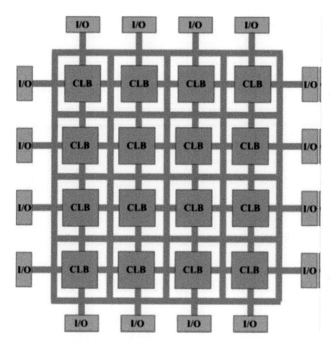

FIGURE 5.1 Generic FPGA architecture.

There are mainly two types of FPGAs:

1. One time programmable (OTP); and
2. SRAM-based reprogrammable.

The difference between these two devices lies in the logic cell implementation interconnection mechanism used inside the device. SRAM-based FPGAs logic cells are programmed by downloading configuration data to the SRAM cells of the CLB look-up table. A system memory or a serial PROM programs the device during the implementation phase.

FPGAs logic cells are connected through anti-fuse elements in a one-time programmable device. They do not need a serial PROM. But once the device is programmed, it cannot be modified.

Details of a CLB are as shown in Figure 5.2. CLB is a basic element in an FPGA which contains the basic logic cells and storage capability for the targeted application design. The CLBs can be configured as a simple logic gate or as an entire processor.

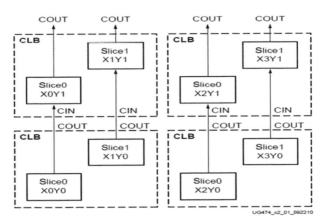

FIGURE 5.2 CLB architecture.

5.2.2 PARTIAL RECONFIGURATION (PR) DESIGN METHODOLOGY

A reconfigurable device (chip) allows the user to change its functionality at any time (including runtime) in the system. Such systems require no manual interventions while changing the functionality or disturbing the existing functionality of digital design. A complete FPGA configuration has to be swapped while performing global reconfiguration, which leads to a change in the internal state of the hardware and thus the FPGA has to restart its operation. This kind of reconfiguration is used for in-field updates. In PR, the user can change the function of part of the FPGA device (usually a region) while other sections remain operational. Furthermore, the PR can be performed either passive (static) by stopping the operation of the application or active (dynamic) where the operation of the application can continue during the reconfiguration.

Optimization of Logical Resources in Reconfigurable Computing

In order to implement any functionality on the FPGA, a specific bit file called configuration file has to be mapped onto the device. A bitstream file will be written in configuration memory. This will be a normal configuration process. But in the PR process, functionality of only selected regions are reconfigured through mapping bit files. Thus PR increases FPGAs computing capabilities. It reduces design size, weight, power, and cost as compared with other conventional design techniques [8]. An operating FPGA design can be modified with the help of partial bit file. After the initial configuration of the device, the functionality of the partially reconfigurable blocks can be altered without disturbing the other regions of the device [9]. A built-in utility component called ICAP helps in modifying the configured data.

Design flow of partially reconfigurable FPGA design is in a way similar to the implementation of multiple non-PR designs when sharing common logic. Partitioning is done to ensure that similar logic between the multiple systems is identical, and its flow can be seen in Figure 5.3.

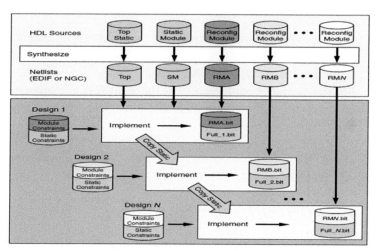

FIGURE 5.3 Partial reconfiguration design methodology.

The grey box at the top represents the synthesis of each module where HDL source are converted into netlists. Full and partial BIT files are generated by implementing appropriate netlists in each design. The static logic obtained after the first implementation is shared commonly among all the subsequent design implementations.

> **Partial Reconfiguration Design Steps:**
> 1. A new project is created in Xilinx Vivado and is set as a PR project.
> 2. The netlist obtained after synthesis from Xilinx Vivado is imported into this project.
> 3. The synthesized netlist shows the different components of the structural implementation of Xilinx Vivado. Some modules of the system are set as reconfigurable partitions (RP).
> 4. Partial blocks (p-blocks) are created for the RPs.
> 5. Reconfigurable modules are added into these partitions. These modules are then linked to the netlists of the individual modules. Faulty and fault free designs are considered.
> 6. Design rules check (DRC) is performed after the creation of P blocks and reconfigurable modules.
> 7. Different configurations are created for indicating faulty and fault free conditions. The netlists are imported and implemented as required.
> 8. These configurations are implemented and promoted to serve as a golden reference for the configurations created in the latter part of the implementation.
> 9. The promoted configurations are verified for implementation errors if any.
> 10. In the final step "Generate Bit Stream" option generates the partial bit files for these configurations. These bit files are generated and are stored in the project folder. These partial bit files can be used to perform reconfiguration at run time.

5.3 CASE STUDY APPLICATIONS

5.3.1 AREA OPTIMIZED ALU

Keeping the main objective of optimizing the utilization of logical resources of ALU, it was designed as a modular system. In this design, different arithmetic operations such as multiplication, addition, subtraction, comparison, and logic gates are been implemented. But to perform PR only multiplication, addition, and logic gates are considered since there are more variants in each type. Comparator and subtraction are considered as static blocks. This partial reconfigurable design is compared with traditional ALU with respect to resource utilization and power [11].

5.3.1.1 DESIGN METHODOLOGY

The design is mainly divided into static and reconfigurable regions. In static region, subtraction, and comparator are implemented and in reconfigurable region multiplication, addition, and logic gates are implemented as shown in Figure 5.4.

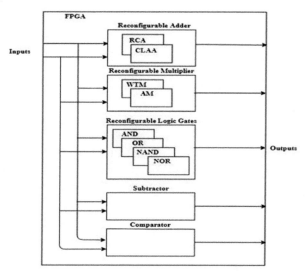

FIGURE 5.4 Block diagram of the design.

Initially, a bit file containing information of full design, i.e., all static and one design in each reconfiguration module is generated and configured on to the FPGA. Thereafter according to the requirement, partial bitstreams for the rest of the configurations are generated and are configured dynamically without altering the main functionality of the design. At any given time, a particular operation in respective reconfigurable module can be implemented without altering the active design.

Figure 5.5 shows the partial definitions in each PR module. Totally, three partition definitions are considered, i.e., PR1 for addition, PR2 for multiplication and PR3 for Logic gates. PR1 has two definitions Ripple adder and Carry look Ahead adder. PR2 has two definitions Wallace tree multiplier and Array multiplier. PR3 has four definitions AND, OR, NAND, and NOR gates. Partial definitions are included in the design with the help of reconfiguration wizard.

FIGURE 5.5 Partial blocks definition.

RTL schematic for ALU without PR can be seen in Figure 5.6 and with PR can be seen in Figure 5.7. Individual functional blocks without using the PR concept are initially synthesized in the Xilinx Vivado tool. Later, a single top module is created based on the PR design method performing all operations.

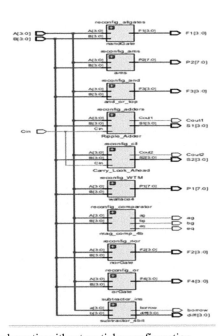

FIGURE 5.6 RTL schematic without partial reconfiguration.

Optimization of Logical Resources in Reconfigurable Computing 79

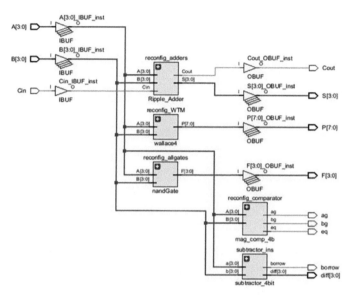

FIGURE 5.7 RTL schematic with partial reconfiguration.

5.3.1.2 IMPLEMENTATION

Figure 5.8 shows the implemented design on the floor plan of the ZedBoard. It can be noticed that PR design is mapped in its specified P-block. Green lines indicate routing.

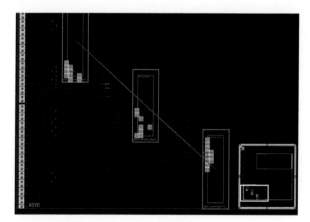

FIGURE 5.8 Floorplan of the implemented design.

5.3.1.3 RESOURCE UTILIZATION REPORT

Figure 5.9 shows the resource utilization report for the design with PR, and Figure 5.10 shows the utilization report for the design without PR. These reports are generated by Xilinx Vivado.

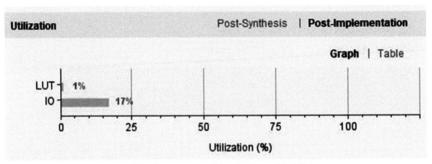

FIGURE 5.9 Resource utilization report of the design with PR.

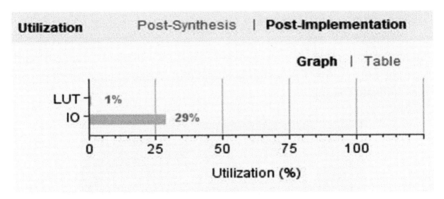

FIGURE 5.10 Resource utilization report of the design without PR.

The resources utilized by PR design are 36 slice LUTs, 15 slices, 36 LUTs and 34 IOBs, and design without PR has utilized 38 slice LUTs, 11 slices, 38 LUTs and 59 IOBs as shown in Table 5.1. It can be observed that design with PR has utilized fewer resources compared with design without PR.

Optimization of Logical Resources in Reconfigurable Computing 81

TABLE 5.1 Utilization Comparison

	Logic Slice LUTs	Logic Slice	LUT as Logic	Bonded IOB
Design without PR	38	11	38	59
Design with PR	36	15	36	34

5.3.2 AREA OPTIMIZED MODULATOR FOR COMMUNICATION SYSTEMS

In communication system, basically there are two types of modulations, i.e., analog, and digital. Now a day's, digital modulation is most popular and is widely used. BPSK, BFSK, QPSK, and QAM are widely used compared with other techniques in digital modulation. A finite number of discrete signals is used to represent digital data in digital modulation. In PSK, a finite number of phases is used to represent unique binary data. In QPSK, four points are mapped on the constellation diagram and two binary bits are used to code a symbol. Compared with BPSK, QPSK can transmit higher data at a given bandwidth. QAM has two components in which one of them is quadrature or Q-Phase component and the other is In-Phase component. QAM is considered as the most efficient modulation technique when compared with other techniques and 16-QAM, 64-QAM and 256-QAM are variants in QAM. More number of constellation points are to be considered to transmit more bits in a symbol [12].

5.3.2.1 DESIGN METHODOLOGY

The main objective of the design is to perform runtime reconfiguration in between different modulation techniques. This design consists of a serial to parallel converter and a reconfigurable module in which BPSK, QPSK, and 16-QAM techniques are reconfigured as shown in Figure 5.11.

Serial to parallel converter here is considered as static block where serial data is converted to parallel data and modulation block is considered as reconfigurable block since reconfiguration is performed only in this block. Initially to configure FPGA, in this case ZedBoard, a bitstream file containing full design, i.e., one static and one modulator (BPSK) block is used. Partial bitstreams files are then used to reconfigure QPSK and

16-QAM dynamically without altering the functionality of the remaining design.

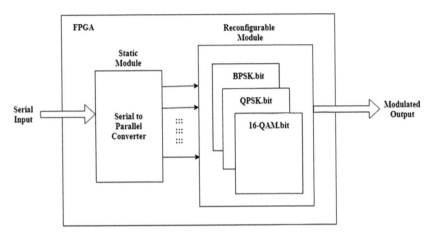

FIGURE 5.11 Block diagram of the modulator.

Figure 5.12 shows the hierarchy of modulator design. Here top module is consisting of static and reconfigurable modules.

FIGURE 5.12 Design hierarchy.

Figure 5.13 shows three partition definitions, i.e., BPSK, QAM, and QPSK defined in one PR region.

FIGURE 5.13 PR definitions.

The RTL schematic of the modulator design is shown in Figure 5.14. It can be observed how static and reconfigurable modules are interconnected. PR_module acts like a black box, and all the three modulation techniques are implemented on at a time depending on the requirements.

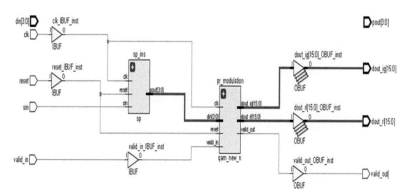

FIGURE 5.14 RTL schematic of the PR design.

5.3.2.2 RESOURCE UTILIZATION REPORT

Table 5.2 shows the utilization of the FPGA resources by the PR design and by the individual modulation blocks.

TABLE 5.2 Resource Utilization Report with Respect to Different Modules

Module Name	Logic Slice Registers	Logic Slice	LUT as Logic	LUT Flop Pairs	Bonded IOB	BufferG Control
BPSK	3	2	1	1	37	1
QPSK	33	16	2	–	40	1
16-QAM	33	18	4	3	40	1
Serial to parallel converter	7	3	–	–	7	1
PR design	13	26	35	4	41	1

5.3.3 RUNTIME RECONFIGURABLE ENCRYPTION ALGORITHMS

Now a day's data security is becoming a tough task since various attacks are making it difficult to communicate data without breach. Encryption plays a vital role in data transfer applications. Ensuring security of data pertaining to basic e-mails and to the most important bank data is the biggest challenge. Normally, in the encryption process, the secret data is combined with a specific key and a ciphertext is generated, which is then communicated. These encryption algorithms are classified into two types: symmetric key and asymmetric key encryption. If the same key is used for encryption and decryption of data, then it is called symmetric otherwise it is asymmetric. AES, TwoFish, 3DES, Blowfish, RC5 algorithms are the important symmetric type algorithms [10].

This design is implemented to validate runtime reconfiguration feature through Encryption algorithms. Two encryption algorithms, namely AES and TwoFish, are designed using Verilog. PR concept is applied for selecting any one algorithm at a time. The proposed design has one RP with two functional reconfigurable modules. The ICAP signals are monitored using an integrated logic analyzer (ILA) and the inputs and outputs are provided using virtual input-output IP (VIO). The entire design is implemented on ZedBoard.

In this design, switching between AES and TwoFish algorithms are implemented using ICAP processor and UART port available on ZedBoard. Initially, individual algorithms are designed, synthesized, and implemented to obtain respective partial bit files. This processor assists in realizing dynamic reconfiguration. Figure 5.15 shows the internal block details of the FPGA present on the ZedBoard. It mainly consists of processor section (PS) and programmable logic section (PL). Initially, in the device floorplan, a single partition (RP) is created to accommodate partial reconfigurable blocks of encryption algorithms.

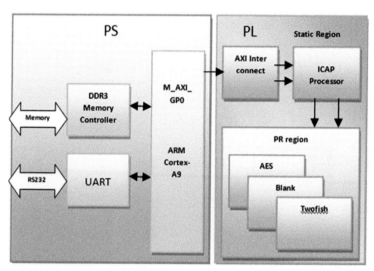

FIGURE 5.15 Block diagram of the design.

This pre-defined partition region gets mapped with any one of the module at a time, depending on the control signal. During the implementation process, AES and TwoFish algorithms are defined individually as reconfigurable modules. Later, AES and TwoFish are added in the RP since both AES and TwoFish modules to be reconfigured. After implementation and bitstream generation, three partial (AES, TwoFish, and Blank) and one full bitstream are generated. Initially, partial bitstream files are stored in SD card, and when PR is initiated, appropriate partial bit files are then stored in DDR, and from there to ICAP processor as shown in Figure 5.15. Reconfiguration is carried out through UART port. Three control inputs are included for specifying a particular encryption module to be executed. They are A, T, and B for the blank module. Three partial bitstreams are reconfigured with a specific name each, i.e., pressing A reconfigures AES; T reconfigures TwoFish and B for Blank. The ICAP signals are monitored using ILA and the inputs and outputs are provided using virtual input/output (VIO). The complete design implementation is done on ZedBoard.

The proposed design is simulated and then synthesized before implementing on the hardware ZedBoard. Figure 5.16 shows the outcome of synthesis process. The synthesis tool generated module-level schematic of the proposed design overview is as in Figure 5.16. Block design named system_i and reconfigurable encoder is highlighted, and the overall design is shown in the right down corner as a world view.

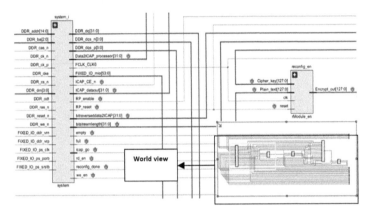

FIGURE 5.16 RTL schematic of the design.

Figure 5.17 shows the floorplan of the implemented design. The highlighted areas show different modules placed on the ZedBoard Zynq Evaluation and Development Kit (xc7z020clg484-1) after placement and routing. PR region in the floorplan is dedicated only for reconfiguration; no static design can be placed in this region.

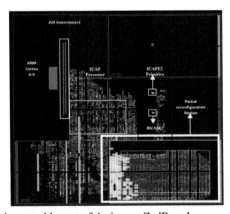

FIGURE 5.17 Implemented layout of design on ZedBoard.

5.4 CONCLUSION

A methodology to execute efficient utilization of logical resources has been illustrated in this chapter. PR technique facilitated in optimizing a number of resources during implementation of multiple functional blocks with the time-multiplexed resources. Three case study applications are considered and explained in detail with respect to logical resource utilization. This chapter mainly highlights the advantages of PR design over traditional FPGA implementation. The same methodology can be applied to any other complex applications.

KEYWORDS

- **advance encryption standard**
- **arithmetic and logic unit**
- **binary amplitude shift keying**
- **configurable logic block**
- **design rules check**
- **field-programmable gate arrays**
- **integrated logic analyzer**

REFERENCES

1. Xilinx. *Partial Reconfiguration User Guide*, April 2018.
2. Mohammed, T. F., Ahmed, H. M., & Ahmed, G. R., (2016). FPGA realization of ALU for mobile GPU. In: *3rd International Conference on Advances in Computational Tools for Engineering Applications (ACTEA)*. 978-1-4673-8523-7/16/31.00, IEEE.
3. Swamynathan, S. M., & Banumathi, V., (2017). Design and analysis of FPGA based 32-bit ALU using reversible gates. *International Conference on Electrical, Instrumentation and Communication Engineering (ICEICE2017)*.
4. Prashanth, B. U. V., Anil, K. P., & Sreenivasulu, G., (2012). Design and implementation of floating-point ALU on a FPGA processor. *International Conference on Computing, Electronics and Electrical Technologies [ICCEET]*.
5. Prasad, B. K. V., & Sai, P., (2016). Implementation of basic digital modulation design models. *Journal of Theoretical and Applied Information Technology, 90*(1).
6. Sanjay, K., Shashi, B., Yogesh, & Jamuna, S., (2019). Design and implementation of two fish encryption algorithm on ZED board. *IJSRR, 8*(5).

7. Xilinx Data Book, (2006). *Field Programmable Gate Arrays*, Xilinx data book user guide published in 2006.
8. Yegireddi, R., & Kumar, R. K., (2016). A survey on conventional encryption algorithms of cryptography. In: *2016 International Conference on ICT in Business Industry and Government (ICTBIG)* (pp. 1–4). Indore. doi: 10.1109/ ICTBIG.2016.7892684.
9. Ye, Y., Yijun, Y., Liji, W., & Xiangmin, Z. (2018). A high performance encryption system based on AES algorithm with novel hardware implementation. *IEEE Conference*. doi: 10.1109/EDSSC.2018.8487056.
10. Yuwen, Z., Hongqi, Z., & Yibao, B., (2013). Study of the AES realization method on the reconfigurable hardware. *2013 International Conference on Computer Sciences and Applications*. doi: 10.1109/CSA.2013.23.
11. Jamuna, S., Dinesha, P., Shashikala, K., & Kishore, K. K. (2019). Area optimized runtime reconfigurable ALU for digital systems. *At International Conference on Advanced Computational and Communication Paradigms (ICACCP) Held at MIT, Sikkim*, February 2019.
12. Jamuna, S., Dinesha, P., Shashikala, K., & Kishore, K. K., (2018). Dynamic reconfigurable modulator for communication systems. *At International Conference on Networking, Embedded and Wireless Systems (ICNEWS-2018)*. BMS College of Engineering, Bangalore.

CHAPTER 6

A SOPHISTICATED SIMILARITY MEASURE FOR PICTURE FUZZY SETS AND THEIR APPLICATION

PALASH DUTTA

Department of Mathematics, Dibrugarh University, Dibrugarh-786004, Assam, India, E-mail: palash.dtt@gmail.com

ABSTRACT

Uncertainty is an unavoidable component of any decision-making process, and fuzzy set theory (FST), as well as the intuitionistic fuzzy set (IFS), are generally explored to deal with it. However, in some complex situations, FST and IFS are not capable of playing a crucial role. In such situations picture fuzzy set (PFS) come into the picture, which is also the direct extension of FST and IFS. Furthermore, similarity measure under an uncertain environment plays an essential role in decision making. In this paper, an attempt has been made to devise similarity measures of PFSs based on the Jaccard Index. Finally, a case study has been carried out under this setting.

6.1 INTRODUCTION

Fuzzy set theory (FST) was first initiated by Zadeh [10] to overcome complexity and deal with uncertainty. FST provides a way to represent value and uncertain concepts by considering partial trustiness, and afterward, many researchers have contributed in generalizing FST. Atanassov [9] developed an intuitionistic fuzzy set (IFS) which deals with partial trustiness degree and partial falseness degree within the same framework. Coung and Kreinovich [1] extended the concept of IFS by incorporating the idea of abstain/neutrality degree in the framework and named as picture fuzzy set (PFS). Afterward, various studies on the properties of PFS were presented [2–5, 8, 11–13, 14–18, 20]. Under uncertain environment, similarity measures (SMs) plays crucial role in decision making for gauging the similarity degree of two objects. Kaufman and Rousseeuw [22] presented some examples to illustrate traditional similarity measure applications in hierarchical cluster analysis. A few numbers of SMs are encountered in literature [6, 7]. However, these approaches are inappropriate to evaluate similarity degree between continuous PFSs. This chapter presents a novel SM for PFSs and applied in a decision-making problem.

6.2 PRELIMINARIES

In this segment, some important notions on FST, IFS, and PFS are presented.

6.2.1 FUZZY SET (FS)

A FS A on X is characterized by its membership function (MF) $\mu_A(x): X \rightarrow [0,1]$ and it is defined as $A = \{(x, \mu_A(x)): x \in X\}$.

6.2.2 IFS

An IFS A on a X is defined as $A = \{(x, \mu_A(x), \nu_A(x)): x \in X\}$, where $\mu_A(x): X \rightarrow [0,1]$ and $\nu_A(x): X \rightarrow [0,1]$ are trustiness degree and falseness degree which obeys $0 \leq \mu_A(x) + \nu_A(x) \leq 1$.

6.2.3 PFS

A PFS A on X is defined as $A = \{(x, \mu_A(x), \eta_A(x), \nu_A(x)): x \in X\}$, where $\mu_A(x): X \rightarrow [0,1]$, $\eta_A(x): X \rightarrow [0,1]$ and $\nu_A(x): X \rightarrow [0,1]$ are trustiness degree, abstain degree and falseness degree, respectively, which obeys $\mu_A(x) + \eta_A(x) + \nu_A(x) \leq 1 \ \forall \ x \in X$.

A Sophisticated Similarity Measure for Picture Fuzzy Sets

6.2.4 ALPHA CUT OF PFS

The α– cut of a PFS A is defines as:

$$C_\alpha(A) = \{x : x \in X \text{ and } \mu_A(x) \geq \alpha, \eta_A(x) \leq \alpha, v_A(x) \leq \alpha\}$$

where; $\alpha \in [0,1]$ and $^\alpha A_+ = \{x \in X : \mu_A(x) \geq \alpha\}$, $^\alpha A_\pm = \{x \in X : \eta_A(x) \leq \alpha\}$ and $^\alpha A_- = \{x \in X : v_A(x) \leq \alpha\}$ are α-cuts of the trustiness MF, abstain MF and falseness MF, respectively of a PFS A.

6.2.5 TRIANGULAR PFSS (TPFS)

A TPFS is denoted by:

$$A = \langle [p_1, q_1, r_1; \beta_1], [p"_1, q_1, r"_1; \gamma_1], [p'_1, q_1, r'_1; \delta_1] \rangle$$ and defined by:

$$\mu_A(x) = \begin{cases} \beta_1 \dfrac{x - p_1}{q_1 - p_1}, & p_1 \leq x \leq q_1 \\ \beta_1 \dfrac{r_1 - x}{r_1 - q_1}, & q_1 \leq x \leq r_1 \\ 0, & \text{Otherwise.} \end{cases}$$

$$\eta_A(x) = \begin{cases} \dfrac{(q_1 - x) + (x - p"_1)\gamma_1}{q_1 - p"_1}, & p"_1 \leq x \leq q_1 \\ \dfrac{(x - q_1) + (r"_1 - x)\gamma_1}{r"_1 - q_1}, & q_1 \leq x \leq r"_1 \\ 1, & \text{otherwise.} \end{cases}$$

$$v_A(x) = \begin{cases} \dfrac{(q_1 - x) + (x - p'_1)\delta_1}{q_1 - p'_1}, & p'_1 \leq x \leq q_1 \\ \dfrac{(x - q_1) + (r'_1 - x)\beta_1}{r'_1 - q_1}, & q_1 \leq x \leq r'_1 \\ 1, & \text{otherwise} \end{cases}$$

6.2.6 TRAPEZOIDAL PFS (TRPFS)

A TrPFS is denoted by:

$$A = \langle [p_1, q_1, r_1, s_1; \beta_1], [p''_1, q_1, r_1, s''_1; \gamma_1], [p'_1, q_1, r_1, s'_1; \delta_1] \rangle \text{ and defined by:}$$

$$\mu_A(x) = \begin{cases} \beta_1 \dfrac{x - p_1}{q_1 - p_1}, & p_1 \leq x \leq q_1 \\ \beta_1, & q_1 \leq x \leq r_1 \\ \beta_1 \dfrac{s_1 - x}{s_1 - r_1}, & r_1 \leq x \leq s_1 \\ 0, & \text{Otherwise.} \end{cases}$$

$$\eta_A(x) = \begin{cases} \dfrac{(q_1 - x) + (x - p''_1)\gamma_1}{q_1 - p''_1}, & p''_1 \leq x \leq q_1 \\ \gamma_1, & q_1 \leq x \leq r_1 \\ \dfrac{(x - r_1) + (s''_1 - x)\gamma_1}{s''_1 - r_1}, & r_1 \leq x \leq s''_1 \\ 1, & \text{otherwise.} \end{cases}$$

$$v_A(x) = \begin{cases} \dfrac{(q_1 - x) + (x - p'_1)\delta_1}{q_1 - p'_1}, & p'_1 \leq x \leq q_1 \\ \delta_1, & q_1 \leq x \leq r_1 \\ \dfrac{(x - r_1) + (s'_1 - x)\delta_1}{s'_1 - r_1}, & r_1 \leq x \leq s'_1 \\ 1, & \text{otherwise} \end{cases}$$

6.3 SIMILARITY MEASURE (SM)

> **Definition 6.1:** If $A = \langle (x, \mu_A(x), \eta_A(x), v_A(x)) \rangle$ and $B = \langle (x, \mu_B(x), \eta_B(x), v_B(x)) \rangle$ are any two PFSs of X, then:
> - $A \subseteq B \Leftrightarrow \mu_A(x) \leq \mu_B(x)$, $\eta_A(x) \geq \eta_B(x)$ and $v_A(x) \geq v_B(x)$, $\forall x \in X$.
> - $A = B$ if $\Leftrightarrow \mu_A(x) = \mu_B(x)$, $\eta_A(x) = \eta_B(x)$ and $v_A(x) = v_B(x)$, $\forall x \in X$.

> **Definition 6.2:** Let us consider a mapping:

$$S: PFS(X) \times PFS() \to [0,1].$$

$S(A,B)$ is called the degree of similarity between the PFSs A and B if in PFS(X) if $S(A,B)$ satisfies the following properties:

- $0 \leq S(A,B) \leq 1;$
- $S(A,B) = 1 \Leftrightarrow A = B;$
- $S(A,B) = S(B,A).$

Now, in Definition 6.3, a novel SM for a PFS A is defined.

➤ **Definition 6.3:** Let $A_1 = \langle [a_1, a_2, a_3, a_4; w], [a_5, a_6, a_7, a_8; h], [a_9, a_{10}, a_{11}, a_{12}; m] \rangle$ and $A_2 = \langle [a'_1, a'_2, a'_3, a'_4; w'], [a'_5, a'_6, a'_7, a'_8; h'], [a'_9, a'_{10}, a'_{11}, a'_{12}; m'] \rangle$ be any two positive trapezoidal PFSs (TrPFS). We choose:

$$\Delta a_i = |a_i - a'_i|, \bar{a}_i = \frac{\min(a_i, a'_i)}{\max(a_i, a'_i)}, \forall i = 1, 2, 3, \ldots, 12$$

$$\Delta w |w - w'|, \bar{w} = \frac{\min(w, w')}{\max(w, w')}$$

$$\Delta h = |h - h'|, \bar{h} = \frac{\min(h, h')}{\max(h, h')}$$

$$\Delta m = |m - m'|, \bar{m} = \frac{\min(m, m')}{\max(m, m')}$$

Then the new similarity measure between the PFSs A_1 and A_2, denoted by, $S(A_1, A_2)$, is defined as:

$$S(A_1, A_2) = \frac{\sum_{i=1}^{12} a_i a'_i + \sum_{i=1}^{12}(1+\Delta a_i)\bar{a}_i + (1+\Delta w)\bar{w} + (1+\Delta h)\bar{h} + (1+\Delta m)\bar{m}}{\sum_{i=1}^{12}(a_i^2 + a'^2_i) - \sum_{i=1}^{12} a_i a'_i + \sum_{i=1}^{12}\{(1+\Delta a_i)^2 + \bar{a}_i^2\} - \sum_{i=1}^{12}(1+\Delta a_i)\bar{a}_i} \quad (1)$$
$$+ (1+\Delta w)^2 + \bar{w}^2 - (1+\Delta w)\bar{w} + (1+\Delta h)^2 + \bar{h}^2 - (1+\Delta h)\bar{h} + (1+\Delta m)^2$$
$$+ \bar{m}^2 - (1+\Delta m)\bar{m}$$

➤ **Theorem 6.1:** $S(A_1, A_2)$ is a similarity measure between the trapezoidal PFSs A_1 and A_2.

> **Proof 6.1:** Since a_i, a'_i, w, h, m are all positive, so $\bar{a}_i, \bar{w}, \bar{h}, \bar{m}$ are all positive and $\Delta a_i, \Delta w, \Delta h$ and Δm are also positive.

$$a_i^2 + a_i'^2 - 2a_i a_i' \geq 0, \forall i = 1, 2, 3, \ldots, 12 \Rightarrow a_i^2 + a_i'^2 - a_i a_i' \geq a_i a_i', \forall i$$
$$= 1, 2, 3, \ldots, 12. \Rightarrow a_i a_i' \leq a_i^2 + a_i'^2 - a_i a_i', \forall i = 1, 2, 3, \ldots, 12. \Rightarrow$$
$$\sum_{i=1}^{12} a_i a_i' \leq \sum_{i=1}^{12} \left(a_i^2 + a_i'^2\right)_i - \sum_{i=1}^{12} a_i a_i'$$

Similarly,

$$\sum_{i=1}^{12} (1+\Delta a_i)\bar{a}_i \leq \sum_{i=1}^{12} \left\{(1+\Delta a_i)^2 + \bar{a}_i^2\right\} - \sum_{i=1}^{12} (1+\Delta a_i)\bar{a}_i, (1+\Delta w)\bar{w} \leq (1+\Delta w)^2$$
$$+ \bar{w}^2 - (1+\Delta w)\bar{w}, (1+\Delta h)\bar{h} \leq (1+\Delta h)^2 + \bar{h}^2 - (1+\Delta h)\bar{h}$$

and,

$$(1+\Delta m)\bar{m} \leq (1+\Delta m)^2 + \bar{m}^2 - (1+\Delta m)\bar{m}$$

Adding,

$$\sum_{i=1}^{12} a_i a'_i + \sum_{i=1}^{12} (1+\Delta a_i)\bar{a}_i + (1+\Delta w)\bar{w} + (1+\Delta h)\bar{h} + (1+\Delta m)\bar{m} \leq \sum_{i=1}^{12} (a_i^2 + a_i'^2) -$$
$$\sum_{i=1}^{12} a_i a'_i + \sum_{i=1}^{12} \{(1+\Delta a_i)^2 + \bar{a}_i^2\} - \sum_{i=1}^{12} (1+\Delta a_i)\bar{a}_i + (1+\Delta w)^2 + \bar{w}^2 - (1+\Delta w)\bar{w} +$$
$$(1+\Delta h)^2 + \bar{h}^2 - (1+\Delta h)\bar{h} + (1+\Delta m)^2 + \bar{m}^2 - (1+\Delta m)\bar{m}$$

Therefore,

$$\left| \frac{\sum_{i=1}^{12} a_i a'_i + \sum_{i=1}^{12} (1+\Delta a_i)\bar{a}_i + (1+\Delta w)\bar{w} + (1+\Delta h)\bar{h} + (1+\Delta m)\bar{m}}{\sum_{i=1}^{12} (a_i^2 + a_i'^2) - \sum_{i=1}^{12} a_i a'_i + \sum_{i=1}^{12} \{(1+\Delta a_i)^2 + \bar{a}_i^2\} -} \right.$$
$$\left. \sum_{i=1}^{12} (1+\Delta a_i)\bar{a}_i + (1+\Delta w)^2 + \bar{w}^2 - (1+\Delta w)\bar{w} + (1+\Delta h)^2 + \right.$$
$$\left. \bar{h}^2 - (1+\Delta h)\bar{h} + (1+\Delta m)^2 + \bar{m}^2 - (1+\Delta m)\bar{m} \right| \leq 1 \Rightarrow S(A_1, A_2) \leq 1$$

and,

$$\left| \frac{\sum_{i=1}^{12} a_i a'_i + \sum_{i=1}^{12} (1+\Delta a_i)\bar{a}_i + (1+\Delta w)\bar{w} + (1+\Delta h)\bar{h} + (1+\Delta m)\bar{m}}{\sum_{i=1}^{12} (a_i^2 + a_i'^2) - \sum_{i=1}^{12} a_i a'_i + \sum_{i=1}^{12} \{(1+\Delta a_i)^2 + \bar{a}_i^2\} - \sum_{i=1}^{12} (1+\Delta a_i)\bar{a}_i +} \right.$$
$$\left. (1+\Delta w)^2 + \bar{w}^2 - (1+\Delta w)\bar{w} + (1+\Delta h)^2 + \bar{h}^2 - (1+\Delta h)\bar{h} + (1+\Delta m)^2 + \right.$$
$$\left. \bar{m}^2 - (1+\Delta m)\bar{m} \right| \geq 0 \Rightarrow S(A_1, A_2) \geq 0$$

Hence, $0 \leq S(A_1, A_2) \leq 1$.

➤ **Proof 6.2:** $A_1 = A_2 \Rightarrow a_i = a'_i, \forall i = 1, 2, 3, \ldots, 12$ and $w = w', h = h', m = m'$

$$\Rightarrow S(A_1, A_2) = \frac{\sum_{i=1}^{12}(a_i^2 + \sum_{i=1}^{12}(1 \times 1) + (1 \times 1) + (1 \times 1) + (1 \times 1)}{2 \times \sum_{i=1}^{12} a_i^2 - \sum_{i=1}^{12} a_i^2 + \sum_{i=1}^{12}(1+1) - \sum_{i=1}^{12}(1 \times 1) +} = \frac{\sum_{i=1}^{12} a_i^2 + 15}{\sum_{i=1}^{12} a_i^2 + 15} = 1$$
$$(1+1) - (1 \times 1) + (1+1) - (1 \times 1) + (1+1) - (1 \times 1)$$

Again,

$$S(A_1, A_2) = 1 \Rightarrow \sum_{i=1}^{12} a_i a'_i + \sum_{i=1}^{12}(1 + \Delta a_i)\bar{a}_i + (1 + \Delta w)\bar{w} + (1 + \Delta h)\bar{h} + (1 + \Delta m)\bar{m} = \sum_{i=1}^{12}(a_i^2 + a'^2_i) - \sum_{i=1}^{12} a_i a_i + \sum_{i=1}^{12}\{(1 + \Delta a_i)^2 + \bar{a}_i^2\} - \sum_{i=1}^{12}(1 + \Delta a_i)\bar{a}_i + (1 + \Delta w)^2 + \bar{w}^2 - (1 + \Delta w)\bar{w} + (1 + \Delta h)^2 + \bar{h}^2 - (1 + \Delta h)\bar{h} + (1 + \Delta m)^2 + \bar{m}^2 - (1 + \Delta m)\bar{m} \Rightarrow \sum_{i=1}^{12}(a_i^2 + a'^2_i) - 2\sum_{i=1}^{12} a_i a'_i + \sum_{i=1}^{12}\{(1 + \Delta a_i)^2 + \bar{a}_i^2\} - 2\sum_{i=1}^{12}(1 + \Delta a_i)\bar{a}_i + (1 + \Delta w)^2 + \bar{w}^2 - 2(1 + \Delta w)\bar{w} + (1 + \Delta h)^2 + \bar{h}^2 - 2(1 + \Delta h)\bar{h} + (1 + \Delta m)^2 + \bar{m}^2 - 2(1 + \Delta m)\bar{m} = 0 \Rightarrow \sum_{i=1}^{12}(a_i - a'_i)^2 + \sum_{i=1}^{12}(1 + \Delta a_i - \bar{a}_i)^2 + (1 + \Delta w - \bar{w})^2 + (1 + \Delta h - \bar{h})^2 + (1 + \Delta m - \bar{m})^2 = 0$$

Therefore,

$\sum_{i=1}^{12}(a_i - a'_i)^2 = 0 \Rightarrow a_i = a_{i'}, \forall i = 1,2,3,\ldots,12 (1 + \Delta w - \bar{w})^2 = 0 \Rightarrow \Delta w = \bar{w} - 1 \Rightarrow \Delta w < 0$ a contradiction, so $\Delta w = 0$

$$\Rightarrow \bar{w} = 1 \Rightarrow w = w'$$

Similarly, $(1 + \Delta h - \bar{h})^2 = 0 \Rightarrow h = h'$ and,

$$(1 + \Delta m - \bar{m})^2 = 0 \Rightarrow m = m'$$

Hence, $A_1 = A_2$
Thus, $S(A_1, A_2) = 1 \Leftrightarrow A_1 = A_2$

➤ **Corollary 6.1:** If A_1 and A_2 are triangular PFSs (TPFS) then also the similarity measure $S(A_1, A_2)$ defined in Eqn. (1) holds good.

6.4 NUMERICAL EXPLANATION

In this section, two examples are illustrated in which similarity is measured.

➤ **Example 6.1:** Suppose the availability of three tea products in the market be presented by the PFSs:

A_1=<[0.3,0.5,0.7,0.9;0.4],[0.2,0.3,0.4,0.5;0.3], [0.1,0.3,0.4,0.6;0.2]>, A_2=<[0,1,2,3;0.7], [0,0.5,0.9,2.5;0.1],[1,1.9,2.3,2.9;0.1]>, and A_3=<[0.35, 0.75,0.99,1.25;0.6],[0.15,0.45,0.65,0.84;0.3],[0.25,0.52,0.88,1;0.3]>.

The availability of the current product dominating the market is:

A=<[0.1,1.9,2,3;0.8],[0.25,0.54,0.77,0.97;0.1],[0.39,0.67,0.95,1.5;0]>

We are to replace the current product by one of the given alternatives.

$$S(A_1, A) = 0.3027$$
$$S(A_2, A) = 0.5930$$
$$S(A_3, A) = 0.3792$$

We have found that A_2 is more similar with A than A_1 and A_3, and hence product A can be replaced by the product A_2, in the market.

➤ **Example 6.2:** Suppose that a panel of experts of Institution prepares a list of journals for possible publication of their researchers and suppose a particular researcher wants to choose a better journal from list of journals in his/her area of interest. Consider that the journals have some weights in the form of TrPFSs Eqn. (7.1) and the ideal weight is given by:

B=<[0,0.5,1,1.5;0.7],[0.20,0.40,0.60,0.90;0.1],[0.37,0.58,0.82,0.9;0.1]>

Finally, the similarity of the journals with the ideal one is evaluated Eqn. (1) using our proposed method (Tables 6.1 and 6.2).

TABLE 6.1 Weights of Journals

Journal-1	<[0.2,0.5,0.8,1;0.6],[0.1,0.6,0.85,1;0.1],[0.3,0.7,0.9,1;0.2]>
Journal-2	<[1,1,1,1;0.9],[1,1,1,1;0],[1,1,1,1;0]>
Journal-3	<[0.66,0.7,0.85,1;0.7],[0.33,0.48,1,1.5;0.1],[0.20,0.45,0.80,1;0.1]>
Journal-4	<[0.1,1.9,2,3;0.8],[0.25,0.54,0.77,0.97;0.1],[0.39,0.67,0.95,1.5;0]>
Journal-5	<[1,2,2,2.5;0.3],[0.9,1,1,1.5;0.5],[0.92,1.1,1.1,1.6;0.1]>

A Sophisticated Similarity Measure for Picture Fuzzy Sets

TABLE 6.2 Similarity Degrees of Papers with B

S (Journal-1, B)	0.7533
S (Journal-2, B)	0.4855
S (Journal-3, B)	0.7180
S (Journal-4, B)	0.5396
S (Journal-5, B)	0.4187

It is observed that the Journal-1 will be chosen by the researcher for possible publication.

6.5 APPLICATION OF THE PROPOSED SM

Consider a panel of experts $D_1, D_2, D_3, \ldots, D_k$ is constituted to determine best product for supplying it in the market among the given n alternatives $A_1, A_2, A_3, \ldots, A_n$ (products) via the m criteria $C_1, C_2, C_3, \ldots, C_m$ for each alternative, respectively. Then, the ideal weights for alternative A_i are chosen the same for each i($i = 1, 2, 3$) as $\tilde{w}$.

The algorithm for the decision making is summarized as:

➢ **Step I:** Experts need to determine linguistic weighting variables (LWVs) for the importance weight of criteria (IWC) and the linguistic rating variables (LRVs) to gauge the ratings of alternatives (RoAs).
➢ **Step II:** Experts gauge IWC via LWVs.
➢ **Step III:** Then, aggregating the weights of criteria (WoC) as:

$$\tilde{w}_j = \frac{1}{K}\left[\tilde{w}_j^1 + \tilde{w}_j^2 + \ldots + \tilde{w}_j^k\right] \tag{2}$$

To obtain the aggregated fuzzy weight (AFW) $\tilde{w}_j$ of the criterion c_j. The new weight vector will be presented as:

$$\tilde{w}_j = \left[\tilde{w}_1' \tilde{w}_2' \ldots \tilde{w}_n'\right]$$

where;

$$\tilde{w}_j' = \left(w_{j1}, w_{j2}, \ldots, w_{jn}\right)$$

> **Step IV:** Experts make their judgement to obtain the aggregated fuzzy ratings (AFR) $\tilde{x}_{ij}$ of alternative A_i for criterion C_j. That is,

$$R = C_1 C_2 \ldots \ldots C_n A_1 \tilde{x}_{11} \tilde{x}_{12} \ldots \ldots \tilde{x}_{1n} A_2 \tilde{x}_{21} \tilde{x}_{22} \ldots \ldots \tilde{x}_{2n} \ldots \ldots A_m \tilde{x}_{m1} \tilde{x}_{m2} \ldots \ldots \tilde{x}_{mn}$$

> **Step V:** Generating the weighted normalized fuzzy decision matrix (WNFDM) as:

$$\tilde{D} = [\tilde{d}_{ij}]_{m \times n}, i = 1, 2, \ldots, m, j = 1, 2, \ldots, n$$

where;

$$\tilde{d}_{ij} = \tilde{r}_{ij} \tilde{w}_j \qquad (3)$$

calculated using basic operations on PFSs [21].

> **Step VI:** Experts gauge $\tilde{d}_i = \sum_{j=1}^{n} \tilde{d}_{ij}$ using basic operations on PFSs.
> **Step VII:** Based on the maximum value of $S(A_i, \tilde{d}_i), i = 1, 2, 3$, experts will choose the best product A_k.

6.5.1 PRACTICAL EXAMPLE

Suppose a panel of experts D_1, D_2 and D_3 is formed to conduct an experiment for selecting the best product for supplying it in the market, among the three products, namely A_1, A_2, and A_3. The ideal weights for alternative A_i are chosen same for each i (i = 1, 2, 3) as $\tilde{w} = \langle [0,1,2,3;0.9], [0,1,2,3;0], [0,1,2,3;0] \rangle$ and five benefit criteria are considered as:

C_1: Availability of raw materials for the product;
C_2: Transportation cost;
C_3: Cost of production;
C_4: Demand for the product; and
C_4: Revenue from the product.

➢ **Step I:** Experts choose the LWVs for the IWC and the LRV to gauge the RoAs with respect to each criterion (Tables 6.3–6.4).

TABLE 6.3 Linguistic Variable for the IWC

Very low (VL)	⟨[0,0,0.1;0.3],[0,0,0.2;0.2],[0,0,0.3;0.4]⟩
Low (L)	⟨[0,0.1,0.25;0.3],[0,0.1,0.3;0.2],[0,0.1,0.35;0.4]⟩
Medium low (ML)	⟨[0.15,0.3,0.45;0.3],[0.1,0.4,0.6;0.2],[0.15,0.24,0.4;0.4]⟩
Medium (M)	⟨[0.3,0.5,0.6;0.3],[0.4,0.6,0.8;0.2],[0.35,0.45,0.55;0.4]⟩
Medium-high (MH)	⟨[0.55,0.7,0.85;0.3],[0.5,0.7,0.9;0.2],[0.6,0.65,0.75;0.4]⟩
High (H)	⟨[0.8,0.9,1;0.3],[0.75,0.9,1;0.2],[0.6,0.7,0.9;0.4]⟩
Very high (VH)	⟨[0.9,0.95,1;0.3],[0.8,0.95,1;0.2],[0.5,0.85,0.95;0.4]⟩

TABLE 6.4 Linguistic Variable for the Ratings

Very poor (VP)	⟨[0,0,0.1;0.3],[0,0,0.12;0.2],[0,0,0.13;0.4]⟩
Poor (P)	⟨[0,0.1,0.2;0.3],[0,0.1,0.25;0.2],[0,0.1,0.35;0.4]⟩
Medium poor (MP)	⟨[0.15,0.2,0.25;0.3],[0.1,0.3,0.6;0.2],[0.15,0.2,0.4;0.4]⟩
Fair (F)	⟨[0.35,0.5,0.65;0.3],[0.45,0.6,0.85;0.2],[0.25,0.55,0.65;0.4]⟩
Medium good (MG)	⟨[0.5,0.7,0.75;0.3],[0.45,0.7,0.9;0.2],[0.3,0.65,0.7;0.4]⟩
Good (G)	⟨[0.6,0.9,1;0.3],[0.65,0.9,1;0.2],[0.6,0.75,0.95;0.4]⟩
Very good (VG)	⟨[0.8,0.9,1;0.3],[0.85,0.9,1;0.2],[0.65,0.7,0.8;0.4]⟩

➢ **Step II:** Experts gauge IWC (Table 6.3) using LWV from Table 6.5.

TABLE 6.5 The IWC Given by Experts

Decision Makers →/Criterion ↓	D_1	D_2	D_3
C_1	VH	VH	MH
C_2	MH	H	VH
C_3	ML	MH	MH
C_4	MH	MH	H
C_5	MH	VH	MH

➢ **Step III:** The WoC are aggregated using Eqn. (2) to obtain the AFW $\tilde{w}_j$ of the criterion C_j and experts provide their opinion (Table 6.2) to obtain the AFRs $\tilde{x}_{ij}$ of A_i using C_j.

$\tilde{w}_1 = =< [0.24,0.86,0.95;0.3],[0.7,0.86,0.97;0.2],[0.53,0.77,0.88;0.4] >$
$\tilde{w}_2 = =< [0.75,0.85,0.95;0.3],[0.68,0.85,0.97;0.2],[0.57,0.73,0.87;0.4] >$
$\tilde{w}_3 = =< [0.42,0.57,0.72;0.3],[0.37,0.6,0.8;0.2],[0.45,0.51,0.63;0.4] >$
$\tilde{w}_4 = =< [0.63,0.77,0.9;0.3],[0.58,0.77,0.93;0.2],[0.6,0.67,0.8;0.4] >$
$\tilde{w}_5 = =< [0.67,0.78,0.9;0.3],[0.6,0.78,0.93;0.2],[0.57,0.72,0.82;0.4] >$

> **Step IV:** The FDM R is evaluated using Table 6.3 and presented in Table 6.6.

As the FDM is normalized itself and so, no need of further normalization.

TABLE 6.6 The Final Aggregate Result Obtained from Ratings Given by Experts

Criterion	Alternative	Linguistic Variable
$^*C_1$	A_1	F
	A_2	MG
	A_3	G
$^*C_2$	A_1	VG
	A_2	MG
	A_3	F
$^*C_3$	A_1	G
	A_2	MG
	A_3	F
$^*C_4$	A_1	F
	A_2	MG
	A_3	G
$^*C_5$	A_1	G
	A_2	G
	A_3	VG

> **Step V:** Then, weighted normalized FDM can be evaluated using Eqn. (3).

➤ **Step VI:** Evaluating $\tilde{d}_i = \sum_{j=1}^{n} \tilde{d}_{ij}$ and $S(\tilde{d}_i, \tilde{w}), i=1, 2,$, (Table 6.7) we have:

TABLE 6.7 Values of $\tilde{d}_i$s and Their Similarity Degrees with $\tilde{w}$

$\tilde{d}_i$	Values of $\tilde{d}_i$	Similarity Degrees $S(\tilde{d}_i, \tilde{w})$
$\tilde{d}_1$	<[1.74,2.56,2.96,3.77;0.3],[1.8,2.84,3.22,4.31;0.3],[1.26,2.2,2.27,3.17;0.2]>	0.5099
$\tilde{d}_2$	<[1.7,2.73,2.9,3.54;0.3],[1.45,2.75,3.03,4.23;0.4],[0.99,2.27,2.3,3.01;0.3]>	0.5268
$\tilde{d}_3$	<[1.8,2.6,3.05,3.84;0.3],[1.83,2.87,3.31,4.33;0.4],[1.25,2.24,2.31,3.24;0.2]>	0.4977

Note: $\tilde{d}_2$ is more similar to the ideal weight $\tilde{w}$, hence the product A_2 will be selected for supplying it in the market.

6.6 CONCLUSION

It is encountered in the literature that there is no similarity measure of continuous PFSs exist. Nevertheless, continuous PFSs are an important tool to deal with complex situation under uncertain environment. Therefore, in this present work done, a maiden effort has been made to devise a similarity measure for continuous PFSs. Afterwards, it is applied in the decision-making problem of supply of goods as well as in other problems. The present model is found to be very effective, logical, valid, and novel.

KEYWORDS

- aggregated fuzzy ratings
- fuzzy set theory
- linguistic rating variables
- similarity measure
- triangular PFSS
- weighted normalized fuzzy decision matrix

REFERENCES

1. Coung, B. C., & Kreinovich, V., (2013). Picture fuzzy set-a new concept for computational intelligence problems. In: *Proceedings of The Third World Congress on Information and Communication Technologies* (pp. 1–6).WIICT.
2. Cuong, B. C., (2014). Picture fuzzy sets. *J. Comput. Sci. Cybern., 30*, 409–420.
3. Cuong, B. C., Kreinovich, V., & Ngan, R. T., (2016). A classification of representable t-norm operators for picture fuzzy sets. In: *2016 Eighth International Conference on Knowledge and Systems Engineering (KSE)* (pp. 19–24). IEEE.
4. Cuong, B. C., & Hai, P. V., (2015). Some fuzzy logic operators for picture fuzzy sets. In: *Seventh International Conference on Knowledge and Systems Engineering* (pp. 132–137).
5. Cuong, B. C., Ngan, R. T., & Hai, B. D., (2015). An involutive picture fuzzy negator on picture fuzzy sets and some de morgan triples. In: *Seventh International Conference on Knowledge and Systems Engineering* (pp. 126–131).
6. Hwang, C. M., Yang, M. S., & Hung, W. L., (2018). New similarity measures of intuitionistic fuzzy sets based on the Jaccard index with its application to clustering. *Int. J. Intell. Syst., 33*(2018), 1672–1688.
7. Wei, G. W., (2017). Some similarity measures for picture fuzzy sets and their applications. *Iran. J. Fuzzy Syst.* https://ijfs.usb.ac.ir/article_3579.html (accessed on 20 April 2021).
8. Garg, H., (2017). Some picture fuzzy aggregation operators and their applications to multicriteria decision-making. *Arab. J. Sci. Eng.*, 1–16. http://dx.doi.org/10.1007/s13369-017-2625-9.
9. Atanassov, K. T., (1986). Intuitionistic fuzzy sets. *Fuzzy Sets and Systems, 20*, 87–96.
10. Zadeh, L. A., (1965). Fuzzy set theory. *Inform. Control, 8*, 338–356.
11. Son, L. H., (2016). Generalized picture distance measure and applications to picture fuzzy clustering. *Appl. Soft Comput. J.* http://dx.doi.org/ 10.1016/j.asoc.2016.05.009.
12. Son, L. H., (2017). Measuring analogousness in picture fuzzy sets: From picture distance measures to picture association measures. *Fuzzy Optim. Decis. Mak.*, 1–20.
13. Son, L. H., Viet, P., & Hai, P., (2017). Picture inference system: A new fuzzy inference system on picture fuzzy set. *Appl. Intell., 46*, 652–669.
14. Dutta, P., & Ganju, S., (2018). Some aspects of picture fuzzy set. *Transactions of A. Razmadze Mathematical Institute, 172*, 164–175.
15. Viet, P. V., Chau, H. T. M., & Hai, P. V., (2015). Some extensions of membership graphs for picture inference systems. In: *2015 Seventh International Conference on Knowledge and Systems Engineering, (KSE)* (pp. 192–197). IEEE.
16. Singh, P., (2015). Correlation coefficients for picture fuzzy sets. *J. Intell. Fuzzy Syst., 28*, 591–604.
17. Phong, P. H., Hieu, D. T., Ngan, R. T. H., & Them, P. T., (2014). Some compositions of picture fuzzy relations. In: *Proceedings of the 7th National Conference on Fundamental and Applied Information Technology Research, FAIRâ€™7* (pp. 19–20). Thai Nguyen.
18. Peng, X., & Dai, J., (2017). Algorithm for picture fuzzy multiple attribute decision making based on new distance measure. *Int. J. Uncertain. Quant., 7*, 177–187.

19. Liang, Z., & Shi, P., (2003). Similarity measures on intuitionistic fuzzy sets. *Pattern Recognit Lett., 24*, 278–285.
20. Dutta, P., (2018). Medical diagnosis based on distance measures between picture fuzzy sets. *International Journal of Fuzzy System Applications (IJFSA), 7*(4), 15–36.
21. Dutta, P., Bora, R., & Dash, S. R., (2019). Operations on picture fuzzy numbers and their application in multi-criteria group decision making problems. In: Dehuri, S., Mishra, B., Mallick, P., Cho, S. B., & Favorskaya, M., (eds.), *Biologically Inspired Techniques in Many-Criteria Decision Making* (Vol. 10). BITMDM-2019, Learning and analytics in intelligent systems, Springer, Cham.
22. Kaufman, L., & Rousseeuw, P. J. (1990). Finding Groups in Data, New York: Wiley.

CHAPTER 7

SEMI-CIRCULAR FUZZY VARIABLE AND ITS PROPERTIES

PALASH DUTTA

Department of Mathematics, Dibrugarh University, Dibrugarh-786004, Assam, India, E-mail: palash.dtt@gmail.com

ABSTRACT

Uncertainty theory is the recently developed technique to model uncertainty, and it is comprised of credibility measures. In this regard for uncertainty modeling, a semi-circular fuzzy variable (SCFV) is presented here, and an attempt has been made to derive the possibility, necessity, and credibility measure of the SCFV. After that, some swot-up are made on expected value, variance, rational upper bound. Numerical illustrations are depicted here to verify the novelty and effectiveness of the SCFV.

7.1 INTRODUCTION

Uncertainty is an integral part of any decision-making process, and no one can avoid it. Uncertainty generally arises because of lack of exactness, insufficient data, Lilliputian sample sizes, fabricated/artificial errors, etc., is an inescapable constituent of the physical world. To handle such types of uncertain situations, a new notion called fuzzy set theory (FST) was established by Zadeh [1]. After that, Zadeh himself [2] developed possibility theory and further studied by minion investigators, e.g., Dubois and Prade [3]; Klir [4]; Yager [5]; Dubois and Prade [6]; Ban [7]; Heilpern [8]; Carlsson and Fuller [9]; and Chen and Tan [10].

Liu and Liu [11] set up a genesis called credibility theory (CT). Further, Li and Liu [12] systematically studied and developed CT. After that some extended studies on CT can be observed in Yi et al. [13]; Garai et al. [14]; and Dutta [15].

Nevertheless, from CT point of view the SCFV is not ventilated even now. Therefore, this Chapter presents a maiden endeavor to study possibility measure (PM), necessity measure (NM) and credibility measure (CM) of the SCFV. Thereafter, a few more properties, e.g., expected value (EV), variance (V), rational upper bound (RUB) of SCFV are discussed. Finally, novelty, and efficiency have been verified through numerical examples.

7.2 PRELIMINARIES

In this segment, a few essential prerequisite notions are presented.

- **Definition 7.1:** For a non-void set X and its power set ρ, a fuzzy variable (FV) is a function $(X, \rho, Pos) \to \mathbb{R}$, where Pos is PM and (X, ρ, Pos) is possibility space [13, 17].
- **Definition 7.2:** For a fuzzy variable A and its membership function (MF) μ and any real number r, the PM of A is defined as [2]:

$$Pos\{A \leq r\} = \sup_{x \leq r} \mu_A(x)$$

- **Definition 7.3:** For a fuzzy variable A and its MF μ and any real number r, the NM of A is defined as [2]:

$$Nec\{A \leq r\} = 1 - \sup_{x > r} \mu_A(x)$$

- **Definition 7.4:** A CM (Cr) the fuzzy variable A is given by its MF μ as:

$$Cr(A \leq r) = \frac{1}{2}\{Sup_{x \leq r}\mu_A(x) + 1 - Sup_{x > r}\mu_A(x)\}, x, y \in \mathbb{R}$$

- **Definition 7.5:** The CD $\Phi_A : \mathbb{R} \to [0,1]$ of a FV A is defined as [16]:

$$\Phi_A(x) = Cr\{\theta \in \Theta : \zeta(\theta) \leq x\}$$

➢ **Definition 7.6:** For a FV A with a regular CD Φ_A, the inverse function Φ_A^{-1} is called the inverse credibility distribution (ICD) of A [18].

7.3 FORMULATION OF SEMI-CIRCULAR FUZZY VARIABLE (SCFV)

It is well known that the common expression of a circle centered at (a,b) is given by:

$$(x-a)^2 + (y-b)^2 = r^2$$

To formulate a normal and convex semi-circular fuzzy variable (SCFV), we take $b = 0$ and $r = 1$ i.e.:

$$(x-a)^2 + y^2 = 1$$

Hence, the most needed MF of the SCFV $A = S_C(a)$ is:

$$\mu_A(x) = \sqrt{1-(x-a)^2},\ a-1 \leq x \leq a+1$$

where; a indicates the mean/core of the FV.

The α-cut of the SCFV $A = S_c(a)$ is:

$$^\alpha A = \left[a - \sqrt{1-\alpha^2}, a + \sqrt{1-\alpha^2} \right]$$

Suppose $A = S_c(15)$ is SCFV acts for an uncertain number. The MF of the SCFV A is:

$$\mu_A(x) = \sqrt{1-(x-15)^2},\ 14 \leq x \leq 16$$

The graphical depiction of A is presented in Figure 7.1.

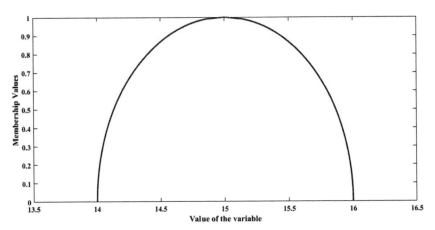

FIGURE 7.1 The SEFV $A = S_c(15)$.

7.4 PM, NM, AND CM OF SCFV

In this segment, an attempt has been made to procure some most important measures such as the PM, NM, and CM of a SCFV.

7.4.1 PMS OF SCFV

For a V $A = S_c(a)$.
The PMs of $(A \leq x)$ and $(A \geq x)$ are defined as:

$$Pos(A \leq x) = \begin{cases} 1, & \text{if } x \geq a, \\ \sqrt{1-(x-a)^2}, & \text{if } x < a, \end{cases}$$

$$Pos(A \geq x) = \begin{cases} 1, & \text{if } x \leq a, \\ \sqrt{1-(x-a)^2}, & \text{if } x > a, \end{cases}$$

The PM of the SCFV $A = S_c(15)$ for $(A \leq x)$ and $(A \geq x)$ are presented in Figures 7.2 and 7.3, respectively.

Semi-Circular Fuzzy Variable and Its Properties

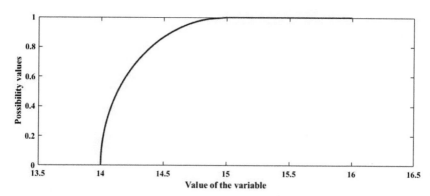

FIGURE 7.2 The PM of SCFV $A = S_c(15)$ for $(A \leq x)$.

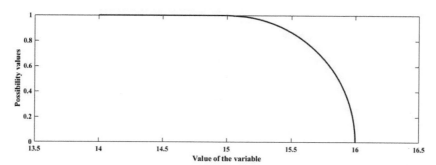

FIGURE 7.3 The PM of SCFV $A = S_c(15)$ for $(A \geq x)$.

7.4.2 NMS OF SCFV

For a V $A = S_c(a)$.

The NMs of $(A \leq x)$ and $(A \geq x)$ are defined as:

$$Nec(A \leq x) = \begin{cases} 0, & \text{if } x \leq a, \\ \sqrt{1-(x-a)^2}, & \text{if } x > a, \end{cases}$$

$$Nec(A \geq x) = \begin{cases} 0, & \text{if } x \geq a, \\ \sqrt{1-(x-a)^2}, & \text{if } x < a, \end{cases}$$

The NM of the SCFV $A = S_c(15)$ for $(A \leq x)$ and $(A \geq x)$ are presented in Figures 7.4 and 7.5, respectively.

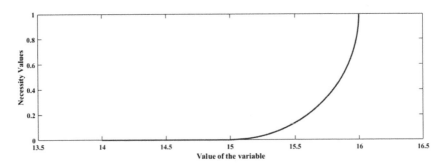

FIGURE 7.4 The NM of SCFV $A = S_c(15)$ for $(A \leq x)$.

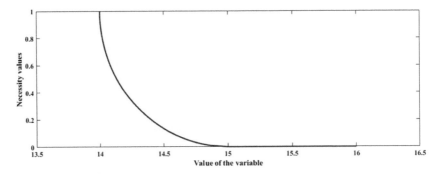

FIGURE 7.5 The NM of SCFV $A = S_c(15)$ for $(A \geq x)$.

7.4.3 CMS OF SCFV

For a SCFV $A = S_c(a)$.

The CM of $A \leq x$ and $A \geq x$ are defined as:

$$Cr(A \leq x) = \begin{cases} \dfrac{1}{2}\sqrt{1-(x-a)^2}, & \text{if } a-1 \leq x \leq a, \\ 1-\dfrac{1}{2}\sqrt{1-(x-a)^2}, & \text{if } a \leq x \leq a+1, \end{cases}$$

Semi-Circular Fuzzy Variable and Its Properties

$$Cr(A \geq x) = \begin{cases} 1 - \frac{1}{2}\sqrt{1-(x-a)^2}, & \text{if } a-1 \leq x \leq a, \\ \frac{1}{2}\sqrt{1-(x-a)^2}, & \text{if } a \leq x \leq a+1, \end{cases}$$

The CMs of the SCFV $A = S_c(15)$ for $(A \leq x)$ and $(A \geq x)$ are presented in Figures 7.6 and 7.7, respectively.

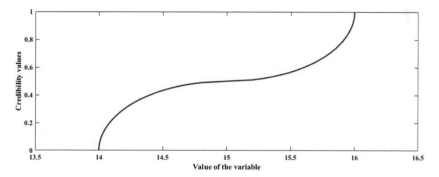

FIGURE 7.6 The CM of SCFV $A = S_c(15)$ for $(A \leq x)$.

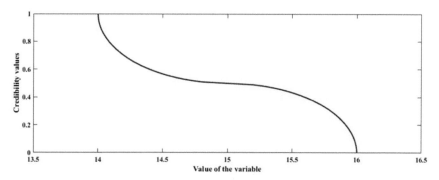

FIGURE 7.7 The CM of SCFV $A = S_c(15)$ for $(A \geq x)$.

7.4.4 CD OF SCFV

As the CD $\Phi_A : \mathbb{R} \to [0,1]$ of a FV A is defined as $\Phi_A(x) = Cr\{\theta \in \Theta : \zeta(\theta) \leq x\}$. Hence, the CD of the SCFV $A = S_c(a)$ is:

$$\Phi_A(x) = \begin{cases} \frac{1}{2}\sqrt{1-(x-a)^2}, & \text{if } a-1 \leq x \leq a, \\ 1 - \frac{1}{2}\sqrt{1-(x-a)^2}, & \text{if } a \leq x \leq a+1, \end{cases}$$

The CD function of the SCFV $A = S_c(15)$ is:

$$\Phi_A(x) = \begin{cases} \frac{1}{2}\sqrt{1-(x-15)^2}, & \text{if } 14 \leq x \leq 15, \\ 1 - \frac{1}{2}\sqrt{1-(x-15)^2}, & \text{if } 15 \leq x \leq 16, \end{cases}$$

7.4.5 ICD OF SCFV

The CD of the SCFV $A = S_c(a)$ is:

$$\Phi^{-1}(\alpha) = \begin{cases} a - \sqrt{1-4\alpha^2}, & \text{if } 0 \leq \alpha \leq 0.5, \\ a + \sqrt{1-4(1-\alpha)^2}, & \text{if } 0.5 \leq \alpha \leq 1, \end{cases}$$

7.5 EXPECTED VALUE (EV)

Liu and Liu [11] provided the EV of FVs using the CD. Zhou et al. [18] presented EV of FVs with the help of ICD.

7.5.1 EXPECTED VALUE (EV) VIA CD

If A is a FV then the EV of A in terms of CD is defined as [11]:

$$E(A) = \int_0^\infty Cr\{A \geq r\}dr - \int_{-\infty}^0 Cr\{A \leq r\}dr$$

Then, the EV of the SCFV $A = S_c(a)$ is:

$$E(A) = \int_0^{a-1} dr + \int_{a-1}^a \frac{1}{2}\sqrt{1-(r-1)^2}\,dr + \int_a^{a+1}\left\{1 - \frac{1}{2}\sqrt{1-(r-1)^2}\right\}dr$$

$$= (a-1) - \frac{1}{2}\int_1^0 \sqrt{t^2}\,dt + 1 - \frac{1}{2}\int_0^1 \sqrt{1-t^2}\,d$$

$$= a + \frac{1}{2}\int_0^1 \sqrt{1-t^2}\,dt - \frac{1}{2}\int_0^1 \sqrt{1-t^2}\,dt$$

$$= a$$

7.5.2 EXPECTED VALUE (EV) VIA ICD

If A is a SCFV then the EV of A in terms of ICD is defined as [18]:

$$E[A] = \int_0^1 \Phi^{-1}(\alpha)\,d\alpha$$

Then, the EV of the SCFV $A = S_c(a)$ is:

$$E(A) = \int_0^{0.5}\left\{a - \sqrt{1-4\alpha^2}\right\}d\alpha + \int_{0.5}^1\left\{a + \sqrt{1-4(1-\alpha)^2}\right\}d\alpha$$

$$= a - \int_0^1 \sqrt{1-t^2}\,dt - \int_1^0 \sqrt{1-t^2}\,dt$$

$$= a - \int_0^1 \sqrt{1-t^2}\,dt + \int_0^1 \sqrt{1-t^2}\,dt$$

$$= a$$

Both the approaches CD and ICD, it is obtained that the EV of the SCFV A is $e = a$.

7.6 VARIANCE

In this part, the variance of SCFV is evaluated via regular CD. Suppose A is a FV with expected value e, then the variance of A is defined as [19]:

$$V[A] = E[(A-e)^2]$$

For the expected value e of the FV which is finite, the variance satisfies:

$$V[A] = E[(A-e)^2] = \int_0^{+\infty} Cr\{(A-e)^2 \geq r\} dr$$

7.6.1 VARIANCE OF A SCFV

To calculate variance $V[A]$ of a SCFV A, the MF of $(A-e)^2$ is required to estimate first and For that α-cut method is used. As the expected valued of SCFV is $e = a$

The α-cut of the SCFV $A = S_c(a)$ is $^\alpha A = \left[a - \sqrt{1-\alpha^2}, a + \sqrt{1-\alpha^2} \right]$.

The procedure is presented below:

$$(^\alpha A - a)^2 = \left[(a - \sqrt{1-\alpha^2} - a)^2, (a + \sqrt{1-\alpha^2} - a)^2 \right]$$

$$= \left[(-\sqrt{1-\alpha^2})^2, (\sqrt{1-\alpha^2})^2 \right]$$

$$= \left[(1-\alpha^2), ((1-\alpha^2) \right]$$

Now, taking $x = (1-\alpha^2)$ gives $\alpha = \sqrt{1-x}, 0 \leq x \leq 1$. Hence, the MF of $(A - e)^2$ is:

$$\mu_{(A-e)^2}(x) = \sqrt{1-x}, 0 \leq x \leq 1$$

As $Cr\{(A-e)^2 < r\} = \dfrac{1}{2} \left\{ \underset{x<r}{Sup}\, \mu_{(A-e)^2}(x) + 1 - \underset{x \geq r}{sup}\, \mu_{(A-e)^2}(x) \right\}$

$$\therefore Cr\{(A-e)^2 < r\} = \frac{1}{2}\{1+1-\sqrt{1-x}\}, 0 \leq x \leq 1.$$

Again, $Cr\{(A-e)^2 \geq r\} = 1 - Cr\{(A-e)^2 < r\}$

Thus, $Cr\{(A-e)^2 \geq r\} = \begin{cases} 1 - Cr\{(A-e)^2 < r\}, 0 \leq x \leq 1 \\ 0, r > 1 \end{cases}$

Then, the variance of SCFV $A = S_c(a)$ is:

$$V[A] = \int_0^\infty Cr\{(A-e)^2 \geq r\}dr.$$

$$= \int_0^1 \frac{1}{2}\sqrt{1-r}\,dr$$

$$= \frac{1}{3}$$

For example, the variance of the SCFV $A = S_c(15)$ is $\frac{1}{3} = 0.33$.

Note: For any SCFV, it is observed that the variance is 0.33.

7.7 RATIONAL UPPER BOUND OF THE VARIANCE (RUBV)

Yi et al. [13] initiated the idea of RUBV.

> **Definition 7.7:** For a FV A with cCD Φ and EV e. The RUBV is defined as [13]:

$$\bar{V}[A] = \int_0^1 \left(1 - \Phi(e+\sqrt{x})\Phi(e-\sqrt{x})\right)$$

> **Definition 7.8:** Let A be a fuzzy variable with credibility distribution Φ. If the expected value is e, then RUBV is evaluated as [13]:

$$\bar{V}[A] = \int_0^1 (\Phi^{-1}(\alpha) - e)^2 d\alpha$$

➢ **Corollary 7.1:** The RUBV of any SCFV is $\frac{2}{3}$
Consider the SCFV $A = S_c(a)$.

$$\int_0^{0.5}(1-4\alpha^2)d\alpha + \int_{0.5}^1 (1-4(1-\alpha)^2)d\alpha$$

$$= 1 + \int_0^{0.5} 4\alpha^2 d\alpha - \int_{0.5}^1 4(1-\alpha)^2 d\alpha$$

$$= 1 - \frac{1}{2}\int_0^1 t^2 dt + \frac{1}{2}\int_1^0 t^2 dt$$

$$= 1 - \int_0^1 t^2 dt$$

$$= 1 - \frac{1}{3}$$

$$= \frac{2}{3}$$

➢ **Corollary 7.2:** Let $A = S_c(a)$ be SCFV. Then, $\bar{V}[A] = 2V[A]$.
Note: For all SCFV, the RUBV is 0.66.
➢ **Corollary 7.3:** Suppose $A = Sc(a_1)$ and $B = S_E(a_2)$ are two SCFV.
Then, $\bar{V}[A+B] = 2(\bar{V}[A] + \bar{V}[B])$.
For the two SCFV A and B, the ICV of $A + B$ is:

$$\Phi^{-1}(\alpha)$$
$$= \Phi_A^{-1}(\alpha) + \Phi_B^{-1}(\alpha)$$
$$= \begin{cases} (a_1 + b_1) - 2\sqrt{1 - 4\alpha^2}, & \alpha \leq 0.5 \\ (a_1 + b_1) + 2\sqrt{1 - 4(1-\alpha)^2}, & \alpha > 0.5 \end{cases}$$

Now:

$$\bar{V}[A+B] = \int_0^{0.5}[(a_1+b_1)-2\sqrt{1-4\alpha^2}\,d\alpha + \int_{0.5}^{1}(a_1+b_1)+2\sqrt{1-4(1-\alpha)^2}\,d\alpha$$

$$= \frac{8}{3}$$

Again, $\bar{V}[A] = \frac{2}{3}$ and $\bar{V}[B] = \frac{2}{3}$.

Since, $\frac{8}{3} = 2\left\{\frac{2}{3} + \frac{2}{3}\right\}$

Consequently, $\bar{V}[A+B] = 2(\bar{V}[A] + \bar{V}[B])$.

> **Corollary 7.4:** Suppose $A = S_E(a_1)$ and $B = S_E(a_2)$ are two SCFV.

Then, $\sqrt{\bar{V}[A+B]} = \sqrt{(\bar{V}[A]} + \sqrt{\bar{V}[B]}$.

Since, $\sqrt{\bar{V}[A+B]} = \sqrt{\frac{8}{3}}$ and $\sqrt{\bar{V}[A]} = \sqrt{\frac{2}{3}}$, $\sqrt{\bar{V}[B]} = \sqrt{\frac{2}{3}}$

Thus, $\sqrt{\bar{V}[A+B]} = \sqrt{(\bar{V}[A]} + \sqrt{\bar{V}[B]}$.

7.8 CONCLUSION

It is experienced that in literature, numerous types of FVs are ventilated, but the most important SCFV has not been investigated yet. Keeping this in mind, for the first time, SCFV has been discussed here via CT. Then, PM, NM, CM, CD, and ICD of SCFV are presented. Moreover, interrogations have been carried out for some properties such as EV of SCFV using CD and ICD. Then, variance and RUVB of SCFV have been examined and established some relationships along with some more properties on RUBV have been discussed for SCFV.

KEYWORDS

- fuzzy set theory
- inverse credibility distribution
- membership function
- necessity measure
- rational upper bound
- semi-circular fuzzy variable

REFERENCES

1. Zadeh, L. A., (1965). Fuzzy sets. *Inform. Control, 8*, 338–356.
2. Zadeh, L. A., (1978). Fuzzy set as a basis for a theory of possibility. *Fuzzy Set and System, 1*, 3–28.
3. Dubois, D., & Prade, H., (1988). *Possibility Theory*. Plenum Press, New York.
4. Klir, J. G., (1992). On fuzzy set interpretation of possibility theory. *Fuzzy Sets Syst., 108*, 263–273.
5. Yager, R. R., (1992). On the specificity of a possibility distribution. *Fuzzy Sets Syst., 50*, 279–292.
6. Dubois, D., & Prade, H., (1987). The mean value of a fuzzy number. *Fuzzy Sets Syst., 24*, 279–300.
7. Ban, J., (1990). Radon-Nikodým theorem and conditional expectation of fuzzy-valued measures and variables. *Fuzzy Sets Syst., 34*(3), 383–392.
8. Heilpern, S., (1992). The expected value of a fuzzy number. *Fuzzy Sets Syst., 47*, 81–86.
9. Carlsson, C., & Fuller, R., (2001). On possibilistic mean and variance of fuzzy number. *Fuzzy Sets Syst., 122*, 315–326.
10. Chen, W., & Tan, S., (2009). On the possibilistic mean value and variance of multiplication of fuzzy numbers. *J. Comput. Appl. Math., 232*(2), 327–334.
11. Liu, B., & Liu, Y. K., (2002). Expected value of fuzzy variable and fuzzy expected value model. *IEEE Trans. Fuzzy Syst., 10*(4), 445–450.
12. Li, X., & Liu, B., (2006). A sufficient and necessary condition for credibility measures. *Int. J. Uncertain. Fuzziness Knowl. Based Syst., 14*(5), 527–535.
13. Yi, X., Miao, Y., Zhou, J., & Wang, Y., (2016). Some novel inequalities for fuzzy variables on the variance and its rational upper bound. *Journal of Inequalities and Applications, 2016*(1), 41, 1–18.
14. Garai, T., Chakraborty, D., & Roy, T. K., (2017). Expected value of exponential fuzzy number and its application to multi-item deterministic inventory model for deteriorating items. *Journal of Uncertainty Analysis and Applications, 5*(1), 1–20.

15. Dutta, P., (2020). Mathematics of uncertainty: An exploration on semi-elliptic fuzzy variable and its properties. *SN Appl. Sci., 2*, 111. https://doi.org/10.1007/s42452-019-1871-8.
16. Liu, B., (2006). A survey of credibility theory. *Fuzzy Optim. Decis. Mak., 5*(4), 387–408.
17. Liu, B., (2004). *Uncertainty Theory: A Branch of Mathematics for Modeling Human Uncertainty.* Springer, Berlin.
18. Zhou, J., Yang, F., & Wang, K., (2015). Fuzzy arithmetic on LR fuzzy numbers with applications to fuzzy programming. *J. Intell. Fuzzy Syst.* doi: 10.3233/IFS-151712.
19. Liu, B., (2007). *Uncertainty Theory.* Springer, Berlin.

CHAPTER 8

VIRTUAL MACHINE SELECTION OPTIMIZATION USING NATURE-INSPIRED ALGORITHMS

R. B. MADHUMALA[1] and HARSHVARDHAN TIWARI[2]

[1]Department of Computer Science Engineering, JAIN (Deemed to be University), Bangalore, Karnataka, India,
E-mail: madhumala8887@gmail.com

[2]CIIRC, Jyothy Institute of Technology, Bangalore, Karnataka, India

ABSTRACT

Cloud computing is the on-demand provisioning of computing resources based on customer requirements. Cloud computing is gaining popularity in the area of processing services with the aid of data centers daily in all domains. Recent trends show that cloud technology is used in every organization results in significantly increased resource utilization in the cloud datacenters. Optimization is very important in the field of science and technology. Virtualization is used to improve resource utilization and to reduce operating costs in a cloud data center. Optimization of virtual machine plays a vital role in improving the resource provision by reducing the total power consumption over the datacenter. Many researchers suggested that nature-inspired algorithms are the best source of optimizing the energy consumption in the datacenter. We are providing a brief introduction of nature-inspired algorithms for virtual machine optimization using particle swarm optimization and ant colony optimization. Particle swarm optimization model is used to solve the multidimensional optimization problems as well as Non-linear optimization problems. Ant colony optimization algorithm is used for continuous domain problems. This chapter provides a survey of nature-inspired algorithms and their applications.

8.1 INTRODUCTION

Cloud computing, considered as a computer paradigm for computing and delivering and services over the internet with many exciting features, provides three major service models like software as a service (SaaS) and infrastructure as a service (IaaS), and platform as a service (PaaS). Consumers always concerned on the performance of the application what they are using and provides are very much interested in providing the efficient resources. The predominant issues in effective cloud service provisioning are optimal resource utilization and the energy conservation over the network. Due to the emerging growth of cloud providers provide IaaS to their main operations. In cloud service models, IaaS model provides virtualized hardware resources where single physical server partitioned into multiple logical servers in turn each logical server act as single physical server for problem execution.

The total energy used and the cost utilization in cloud services mainly depends on the virtual machine scheduling. Efficient resource management through virtual machine placement (VMP) is a great concern in data centers [1]. Allocating resources for cloud data centers to set of virtual machines to the set of physical machines itself is a great challenge. While allocating, there is a need to consider the basic parameters to optimize the resource allocation. We are mainly concentrating on the energy consumptions as well as cost minimization of cloud datacenters.

The cloud service providers provide a group of computational memory and storage units in the form of virtual machines. Each physical bare-metal unit is divided into n number of VM's. The main concern of the cloud provider is to identify a proper physical machine to deploy the VM's. in such a way as to reduce the number of physical machines running while maintaining the service quality. This process is termed as VMP optimization.

The researchers are continuously carried out for effective optimization of VMs. Few researches used heuristic and few used meta-heuristic methods with a flavor of artificial intelligence (AI).

Optimization helps to completely utilize all the resources of a currently running physical machine. Optimization helps in reducing the power consumption, thereby improving the green computing. In real life situations, there will be many constraints for optimizing the cloud center. Maximizing the efficiency of the cloud center and minimizing the undesired factors is really challenging task to solve (Figure 8.1).

Virtual Machine Selection Optimization Using Nature-Inspired Algorithms

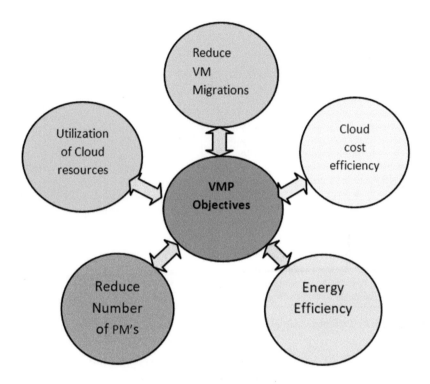

FIGURE 8.1 Virtual machine objectives.

VMP plays an important role in cloud environment. The process is to allocate required resources to virtual machine from physical server. A typical user request will contain specific requirements in terms of CPU, RAM, and Storage. Allocating a virtual machine optimally to a physical server will reduce power consumption, wastage of physical resources, propagation delay, etc. Several approaches are proposed to improve VMP efficiently both in terms of single also as well as multi-dimensional aspects.

Optimization is a trial-and-error method. Many new algorithms will be proposed and their results are tested against the desired metrics, and the successful ones are continuously modified to get the better results. Broadly we can classify the optimization algorithms as conventional and non-conventional algorithms. Much of the work is already done in the conventional algorithms, but very few are proposed in non-conventional algorithms. The non-conventional algorithms are primarily bio- or nature-inspired algorithms. These nature-inspired algorithms need intelligence, and the AI tools are used for inducing the required intelligence into the algorithms.

Few optimization Algorithms are listed under heuristic methods as well as meta-heuristic methods. Optimization Algorithms can be broadly classified as Figure 8.2.

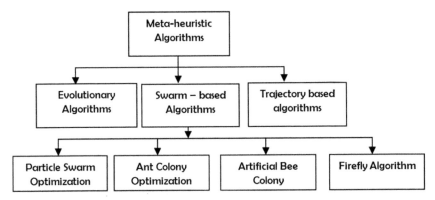

FIGURE 8.2 Optimization algorithms.

In virtual machine selection, overloaded host is detected, further selects VMs to make offload the host mainly to address the performance issue [1]. After a selecting a VM, the host once again checked with the remaining overloaded host. In the next step, it is applied again to select another VM to migrate from the host [2]. The same procedure is repeated until unless the host is considered as being not overloaded.

8.2 MOTIVATION

Virtualization is the key concept of cloud computing. how to select the most suitable host for virtual machine. As stated before, the motivation behind VM placement optimization can be traffic-aware, energy-aware, application-aware, network topology-aware, data aware, and multi-objective [2].

Energy management: minimizing the cost of level at the hardware, server consolidation: green computing. Resource management-resources should available based on the need and cannot be allocated statistically based on the peak workload elasticity of the cloud traffic engineering: to maintain the data enter applications efficiency and also accurate planning of the network architecture. Resources are limited on cloud data centers simply because cloud providers try to minimize overutilized resources. This helps them to have a minimum number of servers as well as reduce the cost for both server hardware and software. Resources like CPU, memory, and hard disk need to be kept minimum to avoid unnecessary costs. So some improvements are needed in its energy cost. We have reviewed recent research on resource provisioning in virtualized computing environments. It reduces response time and maximizes resource utilization, but QoS factors are not considered [3]. Achieves schedule with a lower makespan, but time complexity is more. Minimizes the total cost and risk high in resource management. Existing methods cannot work out the problems of non-coordinated system, such as the solution to the energy field.

8.3 REVIEW OF LITERATURE

This section provides insight into existing algorithms developed which can be used for optimal virtual machine selection and placement in a cloud environment. Aims at scheduling in cloud computing belong to a category of problems known as NP-hard problems due to large solution space. Cloud computing consists of mapping tasks on unlimited computing resources. Nature-inspired algorithms in ant colony optimization (ACO), particle swarm optimization (PSO), and gravitational search algorithm (GSA), which help in solving NP-hard problems, are used in this search. The nature-inspired algorithm in the area of security to protect information associated with tasks and cloud computing:

- Kumar [4] presents a critical survey of nature clever algorithms which are used in AI and automation in real-life domains. Nature-inspired algorithms are emerging area of research on algorithm based on physics and biology [4]. This chapter explains the phenomena of nature to duplicate in artificial systems. Nature-inspired computing and computational intelligence will provide maximum solutions to problems and open new venue in research and development.
- Cloud computing needs optimal resource utilization of cloud resource which intern need a certain novel scheme with enhanced dynamic resource allocation, collective control, resource management, and its maximum distribution in networking and computing

resources. The present trend of using virtualization in resource mobilization for data centers using machine migration techniques. Theja and Babu [5] discusses proposed approaches for optimization of resources for cloud infrastructure [5]. Virtual machines with association with physical machine can be an effective solution for resource optimization by the consideration of certain increased predictive schemes, load balancing and mapping could be increasing factor in virtualization for better performance.

- VM placement in cloud is done by focusing on objects like VM allocation time, energy consumption, SLA violation utilization of resources, etc. [6]. The proposed algorithm that is multi-hybrid ACO-PSO algorithm reduces the resource wastage and power consumption also provides load balancing in servers [6]. It helps in reduction of server costs.
- Usmani and Singh [7] deals with details regarding VM placement algorithm aiming at maximum utilization to reach optimal solution minimization of power consumption. This algorithm aim at studying the workload variability and changing the demands of applications the minimization of trade of between the energy consumption and good performance by using a hybrid technique for server energy efficiency [7]. This is a two-staged process comprising of green computing and overload avoidance.
- Cloud computing has attained a remarkable growth in every field it makes provisioning, scaling, maintenance of applications and serves a breeze. Rani and Bhardwaj [8] focuses on task scheduling ACO genetic algorithm PSO and GSA [8]. This survey deals with task scheduling in cloud computing based on current information and sources to build a good mapping relationship between task and resources. It clearly compares the ACO with other techniques to prove the former as better in comparison.
- Cloud computing, which is an important development in sharing and pooling of resources over internet services, is still in its infancy to achieve improvement much research is required in various directions one is scheduling the goal of scheduling is to trace appropriate resources. This belongs to a category of problem known as NP-hard problem. There are no algorithms which produce optimal solution within polynomial time to solve this problem. Meta heuristic-based techniques provide some

solution within required time. Metaheuristic techniques like ACO GA and PSO and two new techniques like league championship algorithm (LCA) and BAT algorithm (getting inspiration from echolocation behavior of bats yang introduced BAT algorithm) forms these techniques [9]. The comparative analysis of algorithm based on Metaheuristic techniques mainly compares techniques used for Metaheuristic, optimization criteria, nature of tasks, and the environment in which the algorithm is implemented [9].
- Son and Buyya [10] proposes priority aware VM allocation (PAVA), which uses network topology information to allocate VM on the host which is nearest to the requester of the resource. The priority of the task is also considered as parameter [10].

8.4 OPTIMIZATION TECHNIQUES

To maximize or minimize a function so as to seek out the optimum, there are several approaches that one could perform. In spite of a good range of optimization algorithms that would be used, there is not the main one that is considered to be the simplest for any case. One optimization method that is suitable for a drag will not be so for an additional one; it depends on several features, for instance, whether the function is differentiable and its concavity. So as to unravel a drag, one must understand different optimization methods, so this person is in a position to pick the algorithm that most closely fits on the features' problem.

8.4.1 PARTICLE SWARM OPTIMIZATION (PSO)

PSO is an optimization algorithm based on intelligent optimization. Parameters selection of PSO will play an important role in the performance and efficiency of the algorithm. Kennedy and Eberhart [11] proposed the first practical PSO method based on bird flocking. It is metaheuristic optimization algorithm. The basic PSO algorithm works by having a swarm or population of candidate solutions called particles. These particles move around the problem search space as per the simple formula. PSO algorithm may be a method which makes each particle within the population follow the present superior particle by the certain speed, searches problems, optimal solution within the solution space. Particle flying speed is adjusted dynamically by individual and community's flight experience.

Parameters selection of PSO is the key influence on performance and efficiency of the algorithm. It is a complicated optimized problem that the way to determine the optimum parameters for the optimal performance. There are no general methods to work out the optimum parameters, which are selected by user experience, because there are different parameter spaces between different parameters and relativity each other. However, the regularity of various parameter which influences the performance of algorithm could also be found.

Following elements used in PSO:

1. **Particle:** We can define the number of particle in the swarm.
2. **Fitness Function:** It is the function used to find the optimal solution.
3. **Personal Best:** It is the best position of the particle visited so for.
4. **Global Best:** It is the position where the best solution is achieved among all the available particles visited so far.
5. **Velocity Update:** It is determined by the speed and direction of the particle.
6. **Position Update:** All the particles try to move toward the best position to find the optimal solution. Each particle in PSO updates their positions to find the global optimal solution.

In PSO method, each particle tries to modify its current position and velocity according to the distance between the current position of the particle to its pbest position, current position of the particle to its gbest position. Here the below equation shows the modified velocity and position of the swarm movement.

- **New Velocity:** $v_i(k+1) = v_i(k) + c_1 r_1 (pb_i - x_i(k)) + c_2 r_2 (gb - x_i(k))$
- **New Position:** $x_i(k+1) = x_i(k) + v_i(k+1)$

where; i indicates particle index; k indicates time index; v_i: i^{th} particle velocity; x_i: i^{th} particle position; pb_i: personal best position found by i^{th} particle; gb: global best position found by entire swarm (best of pb); c_1, c_2: weighted co-efficient for pbest and gbest; r_1, r_2: uniformly distributed random numbers [0,1].

8.4.2 VARIANTS OF PARTICLE SWARM OPTIMIZATION (PSO)

Discrete PSO algorithm (DPSO) is mainly applicable to discrete as well as combinatorial optimization problems in the given search space. here position of the particles are represented as discrete-valued particles. The velocity and position equation are developed for discrete values and updated in each iteration. during a more general case, when integer solutions (not necessarily 0 or 1) are needed, the optimal solution are often determined by rounding off the important optimum values to the closest integer. Discrete PSO features a very high success rate in solving integer programming problems even when other methods.

Binary PSO (BPSO) algorithm evaluated for continues-valued search spaces, each individual particle of the swarm population and takes a binary decision, either Yes or No, i.e., 1 or 0. Each particle identified by its position as binary values which are 1 or 0. Each particles value later be converted from one to zero or vice-versa.

Multi-objective PSO (MOPSO) The MOPSO algorithm was developed by Coello-Coello et al. two fundamental approaches in this algorithm considered here. In the primary approach, each particle is one objective function at a time, and finding the best out of availability. In the second approach, a particle is evaluated by all the available objective functions, and it is based on the Pareto optimality concept and produces many non-dominated best positions out of that only one particle will be selected to update the particle position and velocity [12].

Perturbed PSO (PPSO) perturbed particle swarm algorithm which is based upon a new particle updating strategy and the concept of perturbed global best (p-gbest) within the swarm. The perturbed global best (p-gbest) updating strategy is based on the concept of possibility measure (PM) to model the lack of information about the true optimality of the gbest [13].

8.4.3 ANT COLONY OPTIMIZATION (ACO) ALGORITHMS

ACO is one of the swarm intelligence (SI) algorithm. ACO is a probabilistic method of finding optimal paths. In computing and researches, the ACO algorithm is employed for solving different computational problems.

ACO was first introduced by Marco Dorigo within the early 90s. This algorithm is introduced based on the foraging behavior of an ant for searching a path between their colony and source food. Initially, it had been wont to solve the well-known traveling salesperson problem (TSP). Later, it is used for solving different hard optimization problems.

Ants are social insects. They sleep in colonies. The behavior of the ants is controlled by the goal of checking out food. While searching, ants roaming around their colonies. An ant repeatedly hops from one place to a different to seek out the food. While moving, it deposits a compound called pheromone on the bottom. Ants communicate with one another via pheromone trails. When an ant finds some amount of food it carries the maximum amount because it can carry. When returning it deposits pheromone on the paths supported the number and quality of the food. Ant can smell pheromone. So, other ants can smell that and follow that path. the upper the pheromone level features a higher probability of selecting that path, and therefore, the more ants follow the trail, the quantity of pheromone also will increase thereon path.

The first ACO algorithm proposed was ant system (AS). AS was applied to some small instances of the TSP with up to 75 cities. On 15 November 1991 Dorigo submitted a manuscript for the ACO which was again revised on 3 September 1993, 2 July 1994, and 28 December 1994 and was finally published IEEE in 1996. it had been an effort by Dorigo to increase traveling salesperson (TSP) problem to the asymmetric traveling salesperson problem (ATSP).

One problem that comes with this algorithm is stagnation. When many ants travel within the problem space they are going to put down tons of pheromones, but as many of them gravitate towards the simplest solution they find, it creates starvation in other parts of the dataset. Though few ants may take other paths, their pheromones will get evaporated eventually making that path as an unreliable path. This suggests that only the trail where many ants travel will have pheromone replenishment and every one other paths are going to be ignored.

Modified and advanced versions of AS are engineered and here we present three such components:

1. **Elitist Ant System:** In Elite AS, they need created specialist ants alongside the traditional ants of AS. In elite AS bonus pheromones are given to the simplest found solution by multiplying the quantity of pheromones given by specialist ants. whenever a best path is found, it will get a huge deposition of pheromones, making it

relevant for extended duration. While the traditional paths degrade over time, the elitist paths having a huge concentration of pheromone stay as valid options for an extended duration, and thus, one elitist path is sufficient to diverge from the present problem space and thus preventing the stagnation.
2. **Rank-Based Ant System**: Ranked AS also use a sort of special ants as just in case of elitist AS, but here they spread the pheromones on several promising paths instead of giving it to single best path. In rank-based ant system, solutions are ranked based on their length. Only few numbers of best ants are allowed to update their paths in the problem space. In ranked AS, if there are fewer specialists than normal ants, then the worst-ranked paths will not get any pheromones?
3. **Min-Max Ant System:** In this system, there are not any specialist ants, it is using only the traditional ants. during this system, there is a cap on minimum and maximum value of pheromone which will be placed on a specific path. Since the algorithm puts a cap on the minimum amount of pheromone to be deposited, a path's pheromone can never drop so low that the trail becomes absolute. Similarly, because the maximum amount of pheromone on a path is fixed a path can never get saturated that it overshadows all other paths. Whenever a 1 or more paths are near minimum and maximum levels, the pheromone levels are smoothed. This promotes the paths with low-level pheromones while still maintaining the standing between the paths. During this method, the pheromones are applied only to the simplest path, ignoring all others. These additions make the essential AS a strong and competitive algorithm.

8.5 CONCLUSION

In this chapter, we discussed about the basic PSO algorithms along with geometrical and mathematical designs of PSO. Each particle's movement and the velocity updated in the search space also updates the acceleration coefficients and particles' neighborhood topologies. During the optimization process, the high speed of convergence sometimes generates a fast loss of diversity which causes undesirable premature convergence. In this chapter, different nature inspired algorithms and its application for optimal selection and placement of virtual machines in a cloud-based environment. The various parameter considered include memory utilization, bandwidth, RAM speed, computation time and cost, etc. Each algorithm has its own limitations and to find a very optimal solution will require developing a new solution based on the specific application requirements.

KEYWORDS

- **ant colony optimization**
- **binary PSO**
- **gravitational search algorithm**
- **league championship algorithm**
- **particle swarm optimization**
- **virtualization**

REFERENCES

1. Zoha, U., & Shailendra, S., (2016). A survey of virtual machine placement techniques in a cloud data center. *Procedia Computer Science, 78*, 491–498.
2. Madhumala, R. B., & Tiwari H. (2020). "Analysis of Virtual Machine Placement and Optimization Using Swarm Intelligence Algorithms". Business Intelligence for Enterprise Internet of Things, EAI/Springer Innovations in Communication and Computing. A study of virtual machine placement optimization in data centers. In: *Proceedings of the 7th International Conference on Cloud Computing and Services Science (CLOSER 2017)*, (pp. 315–322). ISBN: 978-989-758-243-1.
3. Sukhpal, S. G., & Inderveer C., (2016). A survey on resource scheduling in cloud computing: Issues and challenges. *Journal of Grid Computing*.
4. Suresh Kumar, M., Sindhuja, P., & Rama Moorthy, P. (2018). A brief survey on nature inspired algorithms: Clever algorithms for optimization. *Asian Journal of Computer Science and Technology*.
5. Theja, P. R., & Babu, S. K., (2014). Resource optimization for dynamic cloud computing environment: A survey. *International Journal of Applied Engineering Research, 9*(24), 26029–26042.
6. Suseela, B. B. J., & Jeyakrishnan, V., (2014). A multi-objective hybrid ACO-PSO optimization algorithm for virtual machine placement in cloud computing. *Int. J. Res. Eng. Technol., 3*(4), 474–476.

7. Usmani, Z., & Singh, S., (2016). A survey of virtual machine placement techniques in a cloud data center. *Procedia Computer Science, 78*, 491–498.
8. Rani, P., & Bhardwaj, A. K., (2017). A review: Metaheuristic technique in cloud computing. *International Research Journal of Engineering and Technology (IRJET), 4*.
9. Kalra, M., & Singh, S., (2015). A review of metaheuristic scheduling techniques in cloud computing. *Egyptian Informatics Journal, 16*(3), 275–295.
10. Son, J., & Buyya, R., (2019). Priority-aware VM allocation and network bandwidth provisioning in software-defined networking (SDN)-enabled clouds. *IEEE Transactions on Sustainable Computing, 4*(1), 17–28.
11. Kennedy, J., & Eberhart, R., (1995). Particle swarm optimization. *Proceedings of ICNN'95-International Conference on Neural Networks* (Vol. 4, pp. 1942–1948). Perth, WA, Australia. doi: 10.1109/ICNN.1995.488968.
12. Madhumala, R. B., & Harshvardhan Tiwari, (2019). "Virtual Machine Optimization using Nature Inspired Algorithms," *International Journal for Research in Applied Science and Engineering Technology (IJRASET)*, ISSN: 2321–9653.
13. Zhao, X., (2010). A perturbed particle swarm algorithm for numerical optimization. *Applied Soft Computing, 10*, 119–124.
14. Ajith, A., & Amit, K. S. D., (2008). www.softcomputing.net [Online]. http://www.softcomputing.net/aciis.pdf (accessed on 20 February 2021).

CHAPTER 9

EXTRACTIVE TEXT SUMMARIZATION USING CONVOLUTIONAL NEURAL NETWORK

MIHIR, CHANDNI AGARWAL, SWETA AGARWAL, and UDIT KR. CHAKRABORTY

Department of Computer Science and Engineering, Sikkim Manipal Institute of Technology, Sikkim Manipal University, Sikkim, India, E-mail: udit.kc@gmail.com (U. Kr. Chakraborty)

ABSTRACT

Text summarization is a technique of briefing a large text document by extracting its significant information. Extractive text summarization involves direct extraction of sentences from the original document to form a summarized document. The considered task had been an intriguing one for a long time, and thus many approaches had been proposed for the same. This chapter proposes information extraction from a large text document using a convolutional neural network (CNN).

9.1 INTRODUCTION

The amount of information available online today is huge. Moreover, with the spread of the internet, it is now easy to share and avail any information. However, with this ease of availability of information, the problem of information overloading arises. Although data is available in abundance, it is hard to find whether it is reliable or not. Also, not everything in a document is required, a considerably large document might be filled with redundant data and thus only a portion of it contributes to actual information. Therefore, it becomes difficult to obtain the required and authentic information.

Text summarization is a process by which the information content of a large document can be expressed by a comparatively smaller one about the significant information that was conveyed by the original document. It involves both information retrieval and information filtering [1]. Therefore, it can help obtain relevant information as well as time-efficient for tasks that involve a large amount of data.

Text summarization can be further categorized [2] as extractive text summarization and abstractive text summarization. In extractive text summarization, the document is summarized directly by selecting and extracting relevant sentences and adding it to the generated summary. While, in abstractive summarization, the document under consideration is summarized by extracting the information semantically, i.e., to scan the document, understand its meaning, and then the summary is generated according to the semantic knowledge of the document.

Further, the summarization can be done for a single document or multiple documents. Or to say, text summarization can be done for a single document at a time or summarization of text can be performed considering multiple documents. The current chapter proposes an approach of extractive method of text summarization for a single document and presents results obtained thereof.

The proposed method uses a convolutional neural network (CNN)-based technique for extractive summarization of single documents. The document under consideration is converted to vectors embedding semantically meaningful word representations before summarization. The actual summarization uses pre-trained classes for processing by the CNN. Comparisons with blind summaries returned by human readers using BLEU returned good results.

The rest of the chapter has been organized as follows. The following section presents related work done on text summarization with Section 9.3 presenting in brief the various techniques used. Section 9.4 presents the proposed model in detail, and Section 9.5 is dedicated to the experimentation and results. Finally, Section 9.6 concludes the chapter with discussions and future scope of work.

9.2 RELATED WORK

Text summarization had been under consideration for quite a long time, with one of the earliest work done by Luhn in 1958 [3], which extracts statistical information using word frequency and distribution. Luhn approached the problem by marking the sentences according to the sum of the frequency of words in a document and then selecting and extracting the sentences having the highest score.

Since then, various methods of ranking the sentences have been proposed, Edmundson in 1969 [4], proposed a similar ranking technique which along with the previously mentioned criteria of term frequency, also considered pragmatic phrase, words that appeared in the title as well as the location of the sentence in the document.

The increasing availability of information kept text summarization relevant to researchers, and a lot of variety of techniques have been used to approach the problem. Rau et al. in 1989 [5] approached the problem based on linguistic knowledge acquisition and using system for conceptual information summarization, organization, and retrieval technique for text summarization.

Todd Johnson et al. in 2003 [6] suggested generating a summary by creating a word graph based on similarity and removing the redundancy from the document. Also, Rada Mihelcea in 2004 [7] proposed an unsupervised graph-based ranking algorithm for sentence extraction from a text document.

Another popular work of extraction of sentences was proposed by Fattah et al. in 2012 [8] that considered the effect of summary features like the position of a sentence in the document, positive-negative keywords, similarity, and many more. Then the features thus defined were used with various advanced techniques, such as genetic algorithm, mathematical regression to generate feature weights, then neural networks and Gaussian Mixture model to perform summarization. Another similar but comparatively less complex work was proposed by Mendoza et al. in 2014 [9], where similar sentence features were considered along with using genetic algorithms (GAs). The approach as the sentence extraction from text document is a binary optimization problem as assumed fitness to be dependent on various statistical features of the sentence.

Text summarization has also been approached using neural networks by Aakash Sinha in 2018 [10]. Neural networks in combination with other techniques are also been used for the problem under consideration, such as the idea of using neural networks with rhetorical structure theory was proposed by MK AR [11]. Not only combination techniques but also different variations neural networks have been used to approach the problem. Ramesh et al. in 2017 [12] in their work proposed using recurrent neural networks (RNNs) to condense the text document.

One another variation of artificial neural network (ANN) is convolutional neural network (also popularly known as CNN). This chapter proposes a model that incorporates CNN to extract information from a large text document, such that the extracted information alone can convey the significant idea portrayed by the larger text document.

9.3 METHODOLOGIES USED

9.3.1 WORD EMBEDDINGS

Word embeddings are techniques to map textual information to a vector of continuous numbers. Embeddings have been used extensively used to meaningfully represent textual information in numerical space. In the current chapter, the proposed model uses Doc2vec embedding to vectorize the input text before summarization. Proposed by Mikilov et al. [13], Doc2Vec is a refinement over Word2vec embedding [14, 15].

Word embeddings differ from word encoding in retaining word relationships. While an encoding is merely a word representation, embeddings preserve the context and relationships between words. Encodings are inefficient due to volume which may be exceedingly high for large documents.

Word2vec is an embedding technique that offers two novel architectures to convert text to word vectors, viz., Continuous Bag of Words and Skip-gram model, both working on the concept of the local context window. Doc2vec is based on Word2Vec, the difference being that instead of the word being embedded, it considers paragraph vectors. Similar to Word2Vec, Doc2vec also offers two novel architectures, namely, distributed bag of words version of paragraph vector (PV-DBOW) which is similar to skip-gram architecture and distributed memory version of paragraph vector (PV-DM) which is again similar to CBOW model, which is depicted in Figure 9.1.

Extractive Text Summarization Using Convolutional Neural Network

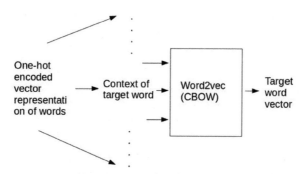

FIGURE 9.1 Rough demonstration of CBOW architecture of Word2Vec.

However, to generate the said distributed representations of word vectors, Doc2Vec takes as input numeric values which are one hot encoded-word representations. So, the input to these embedding architectures are implicitly the encoded vectors and for Doc2Vec, an additional paragraph id is fed as input to uniquely distinguish the paragraphs.

9.3.2 CONVOLUTIONAL NEURAL NETWORK (CNN)

A CNN is a variation of deep neural network having convolution layers, pooling layers, fully connected layers and many more like dropout layers. The input taken is in matrix form, having values that represent the input in its vector form.

A CNN model can be classified into two parts-feature learning and classification. Feature Learning consists of convolutional layers and pooling layers connected. The classification consists of a flattening layer, a fully connected layer and a Softmax function that finally classifies the input having probabilistic values between 0 and 1 (Figure 9.2).

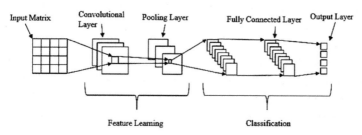

FIGURE 9.2 Schematic diagram of convolution network.

The convolutional layer is the first layer in a CNN model. It takes in two inputs-matrix representation of the input vector and a kernel/filter. The convolutional layer performs Feature Map between the two inputs, which are multiplying the kernel to the input matrix. This helps in highlighting the values that are required for classification. The values of the kernel being used are chosen depending upon the different types of operations like edge detection, blurring, and sharpening. The convolutional layer also uses the ReLU function to bring non-linearity and padding when the kernel does not fit properly in the input matrix.

The pooling layer is used to reduce the number of parameters, i.e., it reduces the dimensionality of the input matrix by retaining important features. Pooling can be done in various ways like max pooling, average pooling, and some pooling.

The fully connected layer works in three parts. Firstly, it performs flattening by converting the matrix obtained after Feature Learning into a vector which is then fed into a fully connected network for creating a model by combining the features. Finally, Softmax function is applied which classifies the input having probabilistic values between 0 and 1 [16].

9.3.3 BLEU SCORE

Bilingual evaluation understudy (BLEU) [17] emerged rapidly as a replacement for humans from evaluating machine-translated text since the human evaluation was expensive and also consumed a lot of time and manpower. BLEU compares a machine-translated text with one or more human-generated reference text and, depending upon the similarity, produces a score. BLEU checks if the words present in the machine-generated text appear in any of the reference text. The degree of similarity in the machine-generated text and the reference text results in a score. The score is computed by dividing the sum of the maximum number of each word present in the reference text to the number of the same word present in the machine-generated text, the score generated is known as unigram precision. The comparison can also be made on unigrams, bigrams, trigrams, 4-grams or n-grams, where these are a group of words with unigrams having 1 word, bigrams having a group of two words and so on. The count of each word present in the text is position-independent. The more the similarity in the number of each word present, the better the machine-translated text and also the precision. The final BLEU score is calculated by taking the geometric mean of the precision scores and then multiplying the computed mean by an exponential brevity penalty factor. The BLEU score computed lies within the range 0 to 1.

This means of evaluation is very similar to calculate the precision of the machine-generated text. The score of 1 is attained very seldomly only when the documents are identical. BLEU also includes "a multiplicative brevity penalty," which becomes active when the target text is longer than the candidate one.

9.4 PROPOSED METHOD

The model proposed in the chapter had been experimented on three datasets, i.e., three documents were tried to be summarized using the proposed model. The documents thus considered had been different from one another in terms of its contents. The documents had information related to World War II [18]; Sourav Ganguly [19]; and Sir Aurobindo [20]. The first document had factual information about the events that occurred in World War II. The second document had information about Sourav Ganguly as a leader, and also his opinions on various events (in first person's speech format). The third document consisted of various quotes of Sir Aurobindo and Mother India.

As no standard datasets were used for the task, data had to be pre-processed for further use as it had lots of noise. Various citations and other garbage symbols had to be removed, a more generalized form of the document was retrieved. Each sentence were then obtained, and it was made sure no sentence were retrieved having number of words less than one, i.e., empty sentences that might appear due to direct retrieval from web pages are filtered. All the documents had to be cleaned and normalized for being fed into the model (Figure 9.3).

The documents had been carefully studied, and various classes of sentences (information-based) had been identified for each of the document. After identification of classes, a dataset had been so prepared that sentences falling under each class had been separated and kept under the heading of identified class. For instance, the document of World War II had sentences having information of events like its advancement, it impact in Asia and Europe, mass slaughter, causalities, aftermath, and so on.

Similarly, all the documents were studied carefully and sentences were classified manually.

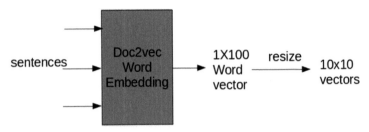

FIGURE 9.3 2D vectorization of sentences.

The chapter proposes a model that uses a neural network to perform the summarization task. But computations cannot be carried out directly over free text, the documents' numerical representations had to be obtained, and for the same, a neural model had to designed just to provide numerical representation of word. For the said task, an embedding scheme had been adopted so that not just word vectors could be obtained, but the vectors even preserved semantic relations amongst them. The embedding technique used for this model was Doc2vec [21] as vectors were not required for each word, but for the entire sentence. Each sentence being treated as individual documents were embedded and equivalent vector representations were retrieved. As the document had sentences ranging from very short sentences to long-running sentences, vector size was fixed to 100, assuming no important information will be lost nor too much void information will be stored. However, as the model proposed in the chapter primarily uses CNN for summarization, the vectors were resized to its 2D equivalent matrix of size 10×10 as the model was seen to perform better with 2D representation as compared to single dimension representation.

The sentence vectors thus obtained are now to be fed into another neural network for training purposes. As said, the documents had been refined, and various classes of sentences were identified based on its information. In accordance to these defined classes, the sentence vectors had to be trained and for this training purpose, a CNN model is proposed. To avoid early shrinking, the model employed a zero-padding layer at the beginning of the architecture, followed by a convolutional layer and max-pooling layer, respectively. But the model uses three parallel architectures for said padding, convolution, and pooling task as the classification task is expected to be better achieved with such architecture [22]. The results of these parallel architectures had to be concatenated to unite the results. After concatenating the results, a flatten layer had been deployed to flatten the outcome of its previous layers removing the insignificant entries. Finally, a dense layer, or to say, a fully connected layer was deployed to complete the classification task.

The model uses rectified linear unit (ReLU) as its activation function in the convolution layer. The activation function gives the same value as input if positive else returns zero. It is said to be one of the most used activation function for CNN architectures [23]. But, this activation function has a limitation that it cannot be used in the output layer. For the output layer, since the network deals with multi-classification task of sentences, the activation of Softmax have been used. The activation function so works that it gives the probability of each sentence for each class and all these probabilities sum to one. To track the loss of the network while training, loss function of sparse categorical cross-entropy [24] is used. The output of this loss function ranges between 0 and 1, and this value is actually in the probability form. As known, loss function is an error reduction method of the network. The generated output is actually the difference between predicted probability and expected probability for the classification task, and thus the correction process aims at depleting this difference.

The architecture of CNN used is a sequential one. However, as the documents are of different type, the architecture had to undergo some changes in order to obtain better results. For instance, since the document of World War II contains simple factual data about the events that occurred during the war period, few changes had to be done for this dataset. Figure 9.4 roughly demonstrates the architecture of World War II data.

However, this is not the case with the other datasets. As document containing the data regarding Sourav Ganguly had first person's speech and that having data related to Sir Aurobindo consisted series of quotes, it had been comparatively complex to summarize the data as compared to the earlier mentioned dataset. Thus for these datasets, an addition pooling layer had to be deployed for better results. Figure 9.5 graphically represents this architecture.

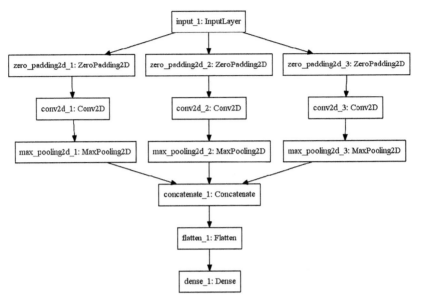

FIGURE 9.4 Rough demonstration of architecture employed for World War II data.

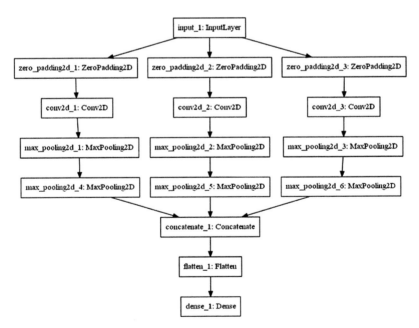

FIGURE 9.5 Rough demonstration of architecture employed for other data.

Using the said architecture, the document is first trained about the sentences falling under specified classes. The network is trained, weights adjusted with each iteration, loss decreased and accuracy is enhanced. Once the network had been trained, the document is given as input, sentences are separated, it is vectorized and then classification is carried out.

The input document is divided into a set of sentences, each classified into designed classes. After classification is done, the first sentences are picked from each class, assuming that the first sentence cannot be skipped for summary generation else the entire context of the class will be ambiguous. The process of selection of sentences is carried on until the summary is so generated that the size of the summary is 30% of that of the size of the original document.

9.5 EXPERIMENT AND RESULTS

The evaluation of the machine-generated summary is not very deterministic task [25]. The correctness of summary may differ from person to person depending upon an individual's own perspective. Yet attempts were made to try evaluating the generated summary (the candidate summary) with reference to some human-generated summary (the reference summary). Four reference summaries were considered for each of the document, and evaluation was carried to check the efficiency of the build model.

For evaluation of the machine-generated summary, the matrix of BLEU has been used. Usually, the BLEU score of less than 15 (if scaled out of 100) is said to be a bad score, while more than 35 [26] is acceptable. The scores obtained for each dataset against each reference summary is given in the tables below.

Table 9.1 presents results that can deduce that reference summary 3 is most similar to that of machine-generated summary, while reference summary 4 is most distinct. Thus it is difficult to say about the correctness of the summary as the same machine-generated summary seems to be good for one human while not that efficient for the other one.

TABLE 9.1 BLEU Score for WW II Dataset

Reference Summary Number	Score
Reference summary 1	0.6
Reference summary 2	0.52
Reference summary 3	0.64
Reference summary 4	0.38

By analyzing Table 9.2, it can be said that if the generated summary is considered in the context of reference summary 3, it is an appreciable model; but again, it is not that good when considered with context of reference summary 4. Also, it can be seen that the results for the first dataset had been quite better than this dataset.

TABLE 9.2 BLEU Score for Sourav Ganguly Dataset

Reference Summary Number	Score
Reference summary 1	0.44
Reference summary 2	0.53
Reference summary 3	0.6
Reference summary 4	0.43

Analysis of data shown in Table 9.3, in reference-to-reference summary 1 the model is seen to provide best result while the same is not the case for reference summary 3. Again, it is clear that the model does not always provide the best results for the same reference set. That implies, for the same human, the summary generated by the model may be really appreciable for one document, but this might not be the case every time.

TABLE 9.3 BLEU Score for Sri Aurobindo Dataset

Reference Summary Number	Score
Reference summary 1	0.61
Reference summary 2	0.54
Reference summary 3	0.45
Reference summary 4	0.49

The considered for evaluation is BLEU. However, BLEU is usually used for the evaluation of machine translation. Therefore, for better understanding for the performance of the designed model, a confusion matrix is created to map the common sentences, its frequency in the candidate summary and the reference summary.

As already mentioned, the summarization task is so carried by the model that it condenses the original document to 30% of its size. Thus reference summaries were also so designed that it included the sentences from the document, and the size ratio was maintained. The documents considered for summarization were condensed to be:

- World War II document – 13 sentences;

- Sourav Ganguly document – 17 sentences;
- Sir Aurobindo dataset – 9 quotes.

The confusion matrix for common sentences in candidate summary and reference summary thus created is shown in Figures 9.6–9.8 for document-related to World War II, Sourav Ganguly and Sir Aurobindo, respectively.

The figures thus show the number of common sentences between the candidate and reference summary as well as amongst the reference summaries themselves.

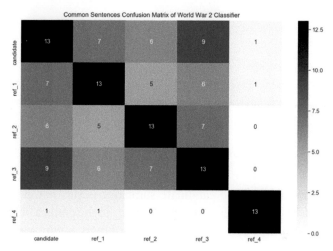

FIGURE 9.6 Confusion matrix for WWII dataset.

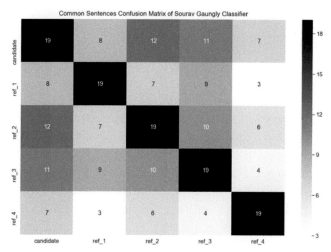

FIGURE 9.7 Confusion matrix for Sourav Ganguly Dataset.

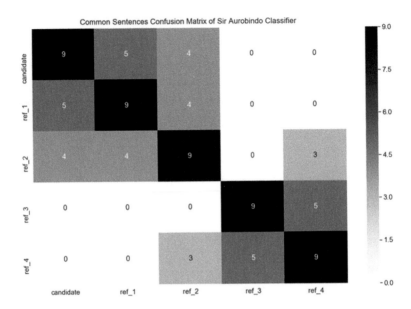

FIGURE 9.8 Confusion matrix for Sri Aurobindo dataset.

But the accuracy of the candidate summary is still not deterministic. Similar to BLEU score, there is no consistency in the pattern of occurrence of common sentences in the candidate and reference summary. Rather, no such consistency can be observed in the reference summaries themselves. Hence, the context of evaluation in such task cannot be fixed, and therefore, it is dependent on an individual whether the summary generated is acceptable or not.

9.6 CONCLUSION

The model proposed in this chapter prepares a summary of text data by using neural network architecture, widely using CNNs. Firstly, the input documented is numerically represented using word embedding technique of Doc2Vec. The vectors from said architecture is then retrieved and fed into another CNN model where the classification task is carried on. After classification of sentences in relevant classes designed in accordance with the document, sentences are retrieved from each class, and the summary is generated, which condenses the original document to 30% of its actual size. The summary thus generated was then evaluated against human generated reference summaries, and it had been found that a simple architecture can generate appreciable results with acceptable computational complexity. However, using CNN makes the model significantly supervised. Thus, for generation of summary using this architecture is possible only when the model can be trained well and this training again involves proper designing of class and rectification of data. But once the model is trained well, it is expected that it can well summarize any document of that domain.

KEYWORDS

- **bilingual evaluation understudy**
- **convolutional neural network**
- **Doc2Vec**
- **embeddings**
- **extractive summarization**
- **linear unit**

REFERENCES

1. Reddy, Y. S., & Siva, K. A. P., (2012). An efficient approach for web document summarization by sentence ranking. *International Journal of Advanced Research in Computer Science and Software Engineering, 2*(7).
2. Prakhar, G., (2019). *Automatic Text Summarization: Simplified.* Towards data science. https://towardsdatascience.com/automatic-text-summarization-simplified-3b7c10c4093a.
3. Luhn, H. P., (1958). The automatic creation of literature abstracts. *IBM Journal of Research and Development, 2(*2), 159–165.
4. Edmundson, H. P., (1969). New methods in automatic extracting. *Journal of the ACM (JACM), 16*(2) 264–285.
5. Rau, L. F., Paul, S. J., & Uri, Z., (1989). Information extraction and text summarization using linguistic knowledge acquisition. *Information Processing and Management, 25*(4), 419–428.
6. Johnson, T., Thede, S., & Vlahov, A., (2003). PARE: An automatic text summarizer. *First Mid-States Conference for Undergraduate Research in Computer Science and Mathematics.*

7. Mihalcea, R., (2004). Graph-based ranking algorithms for sentence extraction, applied to text summarization. *Proceedings of the ACL Interactive Poster and Demonstration Sessions.*
8. Fattah, M. A., & Fuji, R., (2009). GA, MR, FFNN, PNN, and GMM based models for automatic text summarization. *Computer Speech and Language, 23*(1), 126–144.
9. Mendoza, M., et al., (2014). Extractive single-document summarization based on genetic operators and guided local search. *Expert Systems with Applications, 41*(9), 4158–4169.
10. Sinha, A., Abhishek, Y., & Akshay, G., (2018). *Extractive Text Summarization Using Neural Networks.* arXiv preprint arXiv:1802.10137.
11. Kulkarni, A. R., (2015). Text summarization using neural networks and rhetorical structure theory. *International Journal of Advanced Research in Computer and Communication Engineering, 4*(6), 49–52.
12. Nallapati, R., Feifei, Z., & Bowen, Z., (2017). SummaRuNNer: A recurrent neural network-based sequence model for extractive summarization of documents. *Thirty-First AAAI Conference on Artificial Intelligence.*
13. Lau, J. H., & Timothy, B., (2016). *An Empirical Evaluation of doc2vec with Practical Insights into Document Embedding Generation.* arXiv preprint arXiv:1607.05368.
14. Mikolov, T., Chen, K., Corrado, G., & Dean, J., (2013). *Efficient Estimation of Word Representations in Vector Space.* arXiv preprint arXiv:1301.3781.
15. Mikolov, T., Sutskever, I., Chen, K., Corrado, G., & Dean, J., (2013). Distributed representations of words and phrases and their compositionality. *Advances in Neural Information Processing Systems.*
16. Goodfellow, I., Bengio, Y., & Courville, A., (2016). *6.2.2.3 Softmax Units for Multinoulli Output Distributions.* Deep Learning. MIT Press. ISBN.
17. Papineni, K., Salim, R., Todd, W., & Wei-Jing, Z., (2002). BLEU: A method for automatic evaluation of machine translation. *Proceedings of the 40th Annual Meeting on Association for Computational Linguistics.* Association for Computational Linguistics.
18. Wikipedia contributors, (2019). *World War II.* In Wikipedia, the free encyclopedia. Retrieved from: https://en.wikipedia.org/w/index.php?title=World_War_II&oldid=929959693 (accessed on 20 February 2021).
19. Mitter, S., (2018). *Sourav Ganguly: Leadership Lessons from His Memoir.* YSWEEKENDER, https://yourstory.com/weekender/create-a-new-template-for-success-sourav-ganguly (accessed on 20 February 2021).
20. *Sri Aurobindo and the Mother on India.* (2011). Resurgent India. http://new.resurgentindia.org/introduction/sri-aurobindo-and-the-mother-on-india/ (accessed on 20 February 2021).
21. Le, Q., & Tomas, M., (2014). Distributed representations of sentences and documents. *International Conference on Machine Learning.*
22. Kim, Y., (2014). Convolutional neural networks for sentence classification. In: *Conference on Empirical Methods in Natural Language Processing (EMNLP'14).*
23. Viu, D., (2017). *A Practical Guide to ReLU.* Medium. https://medium.com/@danqing/a-practical-guide-to-relu-b83ca804f1f7 (accessed on 20 February 2021).
24. Brownlee, J., (2019). *Loss and Loss Functions for Training Deep Learning Neural Networks.* Machine learning mastery. https://machinelearningmastery.com/

loss-and-loss-functions-for-training-deep-learning-neural-networks/ (accessed on 20 February 2021).
25. Lloret, E., & Manuel, P., (2010). Challenging issues of automatic summarization: Relevance detection and quality-based evaluation. *Informatica, 34*(1).
26. Pan, H. M., (2016). *How BLEU Measures Translation and Why it Matters.* Slator language industry intelligence. https://slator.com/technology/how-bleu-measures-translation-and-why-it-matters/ (accessed on 20 February 2021).

CHAPTER 10

THEORY, CONCEPTS, AND APPLICATIONS OF ARTIFICIAL NEURAL NETWORKS

P. ANIRUDH HEBBAR, M. V. MANOJ KUMAR, and ARCHANA MATHUR

Department of Information Science and Engineering, Nitte Meenakshi Institute of Technology, Bangalore, Karnataka, India, E-mail: manojmv24@gmail.com (M. V. M. Kumar)

ABSTRACT

An artificial neural network (ANN) can be defined as a computational model based on the structure and functions of biological neural networks and find the complex relationship between input and output. The first ANN was invented in 1958 by a psychologist named Frank Rosenblatt. He coined the term as 'perceptron' and visualized it to be like the functionalities of a human brain. It was intended to model how the human brain processed visual data and learnt to recognize objects.

This chapter aims at introducing the building blocks of ANNs and its real-life applications. The content majorly focuses on the structure of biological neuron and how it inspired the discovery of ANNs.

The content of the chapter touches upon various forms of activation functions to decide which neuron should be fired by calculating a weighted sum of inputs and to introduce non-linearity to the output of a neuron. Various widely used backpropagation methods' for optimizing the weights in the hypothesis. The final part of the chapter focuses on discussing the application of ANN in real-life scenarios, such as in healthcare, self-driving vehicles, agriculture, identifying the astronomical objects, etc.

10.1 ARCHITECTURE OF BIOLOGICAL NEURAL NETWORK

Brain is an essential part of human beings which is capable of making everyday decisions. This is possible due to the information the brain collects through different means and uses this information to make sense out of it. It is made up of structural and functional interconnection of what is called neurons. These neurons consist of three vital parts: Synapses, Axons, and Dendrites.

Figure 10.1 depicts a simple neuron. Dendrites collect information from other neurons that are passed through axons and are transmitted to the next neuron via synapse. This happens within a large complex mesh of multiple neurons, thus enables the brain to process information and makes reasonable decisions in an efficient way.

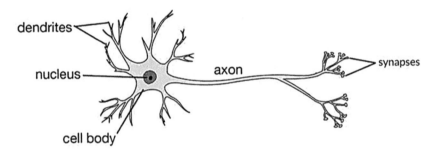

FIGURE 10.1 Image of a single neuron.

Researchers of Free University Amsterdam are successful in understanding the relation between the area of the cortex in the brain and intelligence [1]. Their study shows that people with higher IQ scores have thicker cortex. Thus, a greater number of neurons are present in them. Additional study also shows that when larger dendrites are present, it helps the cells to send the electrical signals faster. Thus, they have the ability to absorb and retain more information at a faster pace.

The researchers are partially successful in trying to capture the very essence of this extraordinary network through machines, by imitating its working and behavior artificially. Thus, it is named artificial neural network (ANN) [2]. The first ANN was invented in 1958 by a psychologist named Frank Rosenblatt. He coined the term as 'perceptron' and visualized it to be similar to the functionalities of a human brain.

Since then researchers have built on top of it and introduced many flavors of the same such as multilayer perceptron (MLP) where its application lies in data compression, ECG noise filtering, etc. Convolutional neural networks (CNN) which works well with image data and are used in applications such as facial recognition, pattern recognition, and other computer vision (CV) applications [3]. Recurrent neural network (RNN) and long short-term memory (LSTM), which is an extension of RNN, are considered to be a sequential model that processes a stream of data over a period of time to generate results. The most popular domains are the Stock Market and the field of natural language processing which involves text data, speech recognition, etc. [4]. This alarms us the potential benefits that ANN and Its variants can offer in solving complex supervised/unsupervised learning tasks in the machine learning (ML) domain.

10.2 ARCHITECTURE OF ANN

ANN is a collection of neurons connected to each other in layers and is capable of processing complex data and solves problems that involve classifying multidimensional objects into classes or generating clusters of different objects. The network of neurons is organized into three distinct layers: input layer, hidden layer, and output layer. Figure 10.2 depicts a two-layered neural network.

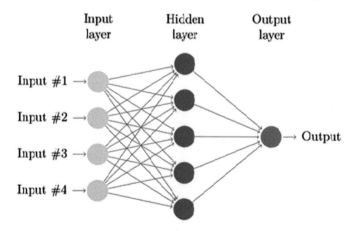

FIGURE 10.2 Image of a simple neural network.

The data comprises of values of attributes/features of different samples. These samples may belong to different classes. The objective is to empower neural network to determine the correct class of new samples by learning behavior of the already classified samples of the training dataset.

In the commonly designed structure of ANN, the neurons of input layers are fully connected with neuron of hidden layers, and a similar pattern is seen across hidden and output layers. The features used in the dataset determine the number of neurons in input layer and similarly, number of classes decides the number of neurons in output layer. Random weights are assigned to the edges connecting all the layers. The hidden nodes and output nodes are equipped with a special function called activation function. In order to understand the flow of data across layers, let us take a sample network (Figure 10.3).

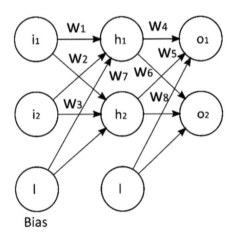

FIGURE 10.3 Sample neural network.

The input layer receives the feature values and the weighted sum of these inputs reaches the hidden layer neurons. The net input at h1 and h2 are given as:

$$h1_{net} = w_1 i_1 + w_2 i_2 + b_1 \tag{1}$$

$$h2_{net} = w_3 i_1 + w_4 i_2 + b_1 \tag{2}$$

Here, $w_1, w_2, w_3,$ and w_4 are the randomly initialized weights and b_1 and b_2 are the bias. Weights are the arbitrarily initialized values to the inputs based on its importance to have a better influence on the output, while bias allows us to adjust the output along with the weighted sum of the inputs. If the activation function used in the network is sigmoid, the squashed output at the hidden nodes is given by:

$$h1_{out} = \frac{1}{1+e^{-h1_{net}}}$$

$$h2_{out} = \frac{1}{1+e^{-h2_{net}}}$$

The flow of data that gets initiated at the input layer and traverses all the way to hidden and later to the output layer is called the *forward propagation*. Similar computation takes place at the output layer:

$$o1_{net} = w_5 h1_{out} + w_6 h2_{out} + b_2$$

$$o2_{net} = w_7 h1_{out} + w_8 h1_{out} + b_2$$

$$o1_{out} = \frac{1}{1+e^{-o1_{net}}}$$

$$o2_{out} = \frac{1}{1+e^{-o2_{net}}}$$

Here, $o1_{out}$ and $o2_{out}$ are the neuron outputs received at the output layer. As soon as the values reaches at the output layer, network tries to calculate error by the standard mean-square-error formula given by:

$$E_{total} = \frac{1}{2} * (t_{01} - o2_{out})^2$$

where; t_{01} is the expected output of the neuron, $o2_{out}$ is the output received from the neuron. During the *backward propagation*, the error gradient is calculated, and the weights are updated. Logically, an error gradient is the indication of the change in error with respect to change in weights. The whole intent is to minimize error and computation of gradient reflects the amount of change in weight that minimizes the error. Let us compute error gradient with respect to w_5, i.e.:

$$\frac{\partial E_{total}}{\partial w_5}$$

$$\frac{\partial E_{total}}{\partial w_5} = \frac{\partial E_{total}}{\partial o1_{out}} * \frac{\partial o1_{out1}}{\partial o1_{net}} * \frac{\partial o1_{net}}{\partial w_5}$$

Solving further, we get:

$$\frac{\partial E_{total}}{\partial w_5} = -(t_{o1} - o1_{out}) * o1_{out} * (1 - o1_{out}) * h1_{out}$$

where; t_{o1} is the target output; $o1_{out}$ is the output from neuron of output layer; $h1_{out}$ is the output from neuron at hidden (previous) layer.

Once $\frac{\partial E_{total}}{\partial w_5}$ is computed at output layer neurons, the weights are updated as indicated below (for w_5):

$$w_5 = w_5 + \varsigma * \frac{\partial E_{total}}{\partial w_5}$$

where; w_5 is the weight, η is the learning rate, and $\frac{\partial E_{total}}{\partial w_5}$ is the error gradient.

10.3 CHARACTERISTICS OF ARTIFICIAL NEURAL NETWORKS (ANNS)

Neural networks offer an array of functions due to its inherent properties; some of the most significant characteristics of ANN are as follows:

1. Neural networks offer an array of functions due to its inherent properties, some of the most significant characteristics of ANN are.

2. Neural networks can train by examples. ANN can be trained with data of known class labels and identify the patterns, before they are made to predict with a new set of data but without the labels.
3. Neural networks can map the input patterns with output patterns.
4. Neural networks can generalize. This means that a neural network can predict with a new set of data based on the past trends.
5. Neural networks are a robust and fault-tolerant model. It can perform with ease irrespective of the incomplete, noisy patterns in the data.
6. The collective behavior of neurons defines its computational power, and no single neuron carries specific information.

The sigmoid activation function is used for calculating the output of each neuron in the network (in Figure 10.3); similarly, there exist an array of activation functions which can be used to perform similar actions under varying requirements. Some widely used activation functions are discussed in the upcoming section.

10.4 ACTIVATION FUNCTION

Activation functions are used to introduce non-linearity into the input which is sent by the input node. It means they decide if a neuron in the hidden layer should be fired or not based on the relevance of the input it receives. Activation functions are required to perform complex operations both on image data as well as text data as the input. These are also required to update weights and bias of the neuron during the back-propagation step. This will be updated in the upcoming section.

Now that we understand what activation function is, let us now discuss various activation functions which can be suitable to your model.

10.4.1 SIGMOID ACTIVATION FUNCTION

This activation function is generally used when dealing with binary classification model. The range of this function is between 0 and 1 range. If the value from the sigmoid function is greater than 0.5, it can be assumed as 1, or else 0. Equation for this activation function is given by:

$$y(x) = \frac{1}{1+e^{-x}}$$

where; x is the input feature passed onto the function (Figure 10.4).

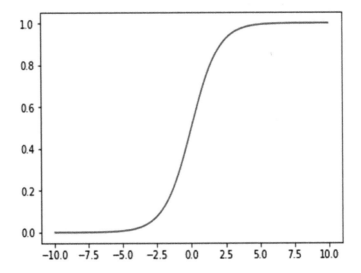

FIGURE 10.4 Sigmoid activation function (activation vs. output).

Following is the implementation of sigmoid activation function. The sigmoid function multiplies with the weighted average of the input in the hidden layers before propagating it to the output layer, whereas the sigmoid derivative is used to find the error rate, i.e., by finding the derivative of the current hidden layer which is then passed onto the previous layer.

Given below are the Python implementation of sigmoid activation function and its derivative:

def sigmoid (self, X)

$$\frac{1}{1+np.exp(-X)}$$

def sigmoid_derivative(self,X):

$$return\ X*(1-X)$$

10.4.2 TANH ACTIVATION FUNCTION

Tanh activation function is a better version of the sigmoid activation function, since it has a better range (−1 to 1), thus centering the data by bringing the mean close to zero, which makes it easier for further layers to learn. Equation for this activation function is given by (Figure 10.5):

$$y(x) = \frac{e^x - e^{-x}}{e^x + e^{-x}}$$

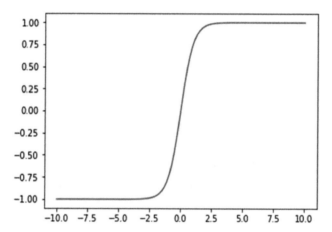

FIGURE 10.5 Tanh activation function.

Following is the implementation of Tanh activation function. The Tanh function multiplies with the weighted average of the input in the hidden layers before propagating it to the output layer, whereas the Tanh derivative function is used to find the error rate in the current hidden layer which is then passed onto the previous hidden layer.

Given below are the python implementations of Tanh activation function along with its derivative:

def tanh(self,X):
return np.tanh (X)
def tanh_derivative(self,X):
return 1 − *np.tanh*(*X*) * *2

10.4.3 RECTIFIED LINEAR UNIT (RELU)

This activation function is the most widely used function and commonly seen in CNN. This activation function is popular since it is computationally less expensive hence faster than Sigmoid as well as Tanh. This is because ReLU involves simpler mathematical operations, thus activating the neurons sparsely. Equation for this activation function is given by (Figure 10.6):

$$F(x) = max(0,x)$$

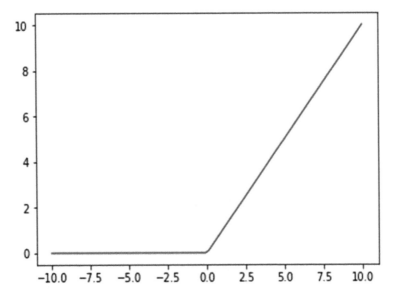

FIGURE 10.6 ReLU activation function.

Following is the implementation of ReLU activation function. The ReLU function multiplies with the weighted average of the input in the hidden layers before propagating it to the output layer, whereas the ReLU derivative is used to find the error rate, i.e., by finding the derivative of the current hidden layer which is then passed onto the previous layer.

Below is the python implementation of ReLU activation function along with its derivative:

```
def ReLu(self, X)
  return X * (X > 0)
def ReLu_derivative(self, X):
  return * (X > 0)
```

10.4.4 SOFTMAX ACTIVATION FUNCTION

This activation function is used in the output layer, unlike other functions which are used in the hidden layers. Although this function is like sigmoid function, the functionalities between the two differ. Consider a network with two output nodes and an input has been passed into the network. The output nodes predict the class and give us a prediction value. When Softmax activation function is applied, the predicted values are reduced to the probability values (between 0 and 1), where the sum of the probability is equal to 1. This gives us the classification error made by the model, which is then backpropagated to the network and weights are updated accordingly. Equation for this activation function is given by (Figure 10.7):

$$P\left(y = j\Theta^{(i)}\right) = \frac{e^{\Theta^{(i)}}}{\sum_{j=0}^{k} e^{\Theta_k^{(i)}}}$$

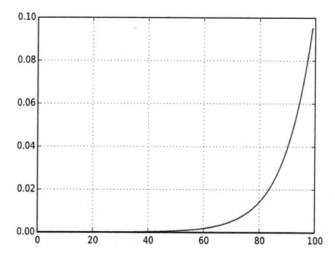

FIGURE 10.7 Softmax activation function.

10.4.5 SAHA BORA ACTIVATION FUNCTION (SBAF)

Most of the above-mentioned functions (sigmoid, tanh) may work well, if the feature space is finite-dimensional and relatively less complex. As feature space grows more complex, neural networks may face a difficult time getting themselves trained within a finite time duration. The chances of attaining a global maximum become weak, and the network may get stuck in local optima. Apparently, the problem of getting stuck in local oscillations can be resolved by fine-tuning the learning rate, but this can become a time-consuming task and thereby, a need to introduce a new activation function becomes evident. SBAF is an activation function that can resolve the problem of attaining global optima in finite time, because of its mathematical structure given by:

$$y = \frac{1}{\left(1 + kx\alpha(1-x)(1-\alpha)\right)}$$

where; α is an exponent, k is a coefficient whose values range between 0 and 1, x is the input and y is the output. The derivative of SBAF can be derived as:

$$\frac{dy}{dx} = \frac{y \times (y-1)}{x \times (x-1)} \times (\alpha - x)$$

This derivative is needed to compute the error gradient for updating the network's weights during backpropagation. It is important to observe that the SBAF derivative involves 'x,' the input to the neuron, in its computation. Figure 10.8 reflects the surface plot of SBAF where x-axis is x, y-axis is α, values of x and α varies between 0.0 and 1.0 keeping k as 1.

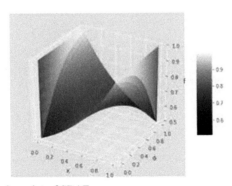

FIGURE 10.8 Surface plot of SBAF.

def sbaf(act)

α = 0.5

$$y = \frac{1.0}{\left(1.0 + \left(k * act^{á} * (1 - act)^{á-}\right)\right)}$$

y = abs(y)

return y

def sbaf_derivative(output, inp):

$$derv = \frac{output * (1.0 - output)}{inp * (1.0 - inp)}$$

derv = derv * (inp − 0.5)

return derv

The above code shows the implementation of SBAF in python. The function parameter 'act' is a linear combination of inputs (i_1, i_2...) and weights (w_1, w_2...). The values of parameters alpha α and k are fixed to 0.5 and 0.91, respectively. A point to note is that the function returns the absolute value of y (i.e., abs(y)). This is because the output 'y' of SBAF becomes a complex number for some values of 'act' (precisely, for negative values of 'act'). Consequently, absolute values are used to remove the effect of the imaginary part. One may argue that this could impact the classification accuracy in some way, the answer to which is-no, it cannot! Another important point worth noticing is the derivative function uses the 'input to neurons' as one of the parameters. This comes from the fact that the derivative of SBAF uses 'x,' which is a variable that stores net-input to neurons, at hidden as well as output layers. Evidently, one has to use a mechanism to store net-input to every neuron during forward propagation. This is unlike the classical activation functions (sigmoid, tanh, RELU, etc.), whose derivatives do not use 'x' at all.

Now that we have discussed how different activation functions, we will now see how Weights are updated in order to improve the performance of the model.

10.5 FORWARD AND BACKWARD PROPAGATION

Though this has been covered in the earlier section, let us summarize the steps once again. Consider a neural network has been initialized for solving some classification problems. The initial step is known as Forward Propagation, where we initialize random weights and biases of the network and propagate it across all hidden layers and output layer. Inwardly, the weighted sum of the inputs (denoted by Z) is calculated at every neuron, which is given as:

$$Z = sum(weights * inputs) + bias$$

Weights are then randomly assigned values to the edges of the network. Input is the attribute values of the dataset.

After calculating Z, activation function is applied to it that squashes the value of Z to range between 0 and 1 (in case sigmoid is used). The squashed input is propagated further in the network and the output layer generates the final prediction of the class label for that specific sample. If the prediction is far-off compared to the actual output, then the cost value (error rate in the prediction) is calculated, which is then backpropagated back to the input nodes where the weights and biases are adjusted again and is once again propagated back to the network. This step is called backpropagation [5]. The cycle repeats until the error rate is reduced. This step thus helps us in improving the performance of the model.

10.6 OPTIMIZATION ALGORITHMS

In the previous section we had discussed about calculating the cost value in the backpropagation step. This step is popularly called Gradient Descent. These cost functions or the optimization algorithms help us in building robust models, by reducing the error rate in our prediction.

There are many algorithms that help us achieve this goal, but we are going to discuss ones which are more widely used. They are:

- Batch gradient descent;
- Mini-batch gradient descent;
- Stochastic gradient descent.

10.6.1 BATCH GRADIENT DESCENT

In this type of algorithm, we consider an entire dataset as a sample and then calculate the sum of the errors for every epoch (iteration over the dataset). The main advantage while implementing this algorithm is that a fixed learning rate can be set, thus the learning can happen at a constant rate. Also, since we are considering the entire dataset for training the fluctuations while calculating the minima is quite less (Figure 10.9).

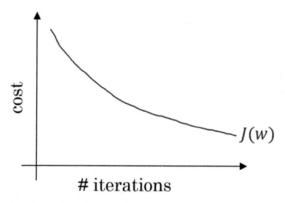

FIGURE 10.9 Batch gradient descent (iterations vs. cost).

The drawback which can be thought of, for this algorithm is that the computation can be heavy if the dataset is too large. While this can work effectively for smaller datasets, the time required to calculate the sum for larger datasets is more. Also, since it is iterating over the dataset repeatedly, the learning curve is not great for the model, thus the contribution to the overall update in weights are negligible.

10.6.2 MINI-BATCH GRADIENT DESCENT

In this type of gradient descent, instead of iterating over the entire dataset to find the cost value, it randomly picks small batches within the dataset and then computes the cost. Thus, the computation to calculate and update the cost is less compared to batch gradient descent. Also, the learning curve for the model to update the weights is better compared to batch gradient descent since the samples are picked up at random, although there can be certain presence of noise with respect to the path taken to reach minima due to the same reason (Figure 10.10).

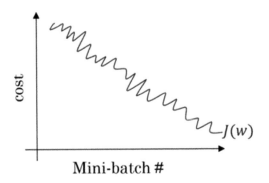

FIGURE 10.10 Mini-batch gradient descent.

10.6.3 STOCHASTIC GRADIENT DESCENT

The word 'Stochastic' in Stochastic Gradient Descent means randomness. It means that update happens for one record at a time instead of the entire dataset. That is, we find out the gradient of a single sample for each epoch instead of the sum of the gradient of the cost function like we saw in batch gradient descent. This becomes handy when you are trying to train a neural network with a large dataset since the computation required to train is comparatively low. The only setback which can be considered is that since the samples are taken at random, the path taken to reach the minima can be random and noisier, compared to the usual gradient descent algorithm (Figure 10.11 and Table 10.1).

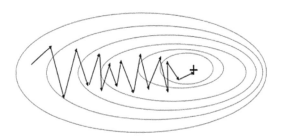

FIGURE 10.11 Path taken stochastic gradient to reach the global minima.

TABLE 10.1 Comparison between Different Gradient Descent Algorithms

	Batch Gradient Descent	Minibatch Gradient Descent	Stochastic Gradient Descent
Accuracy	Good	Good	Good (with annealing)
Update speed	Slow	Medium	High
Memory usage	High	Medium	Low
Time	Slow	Medium	High
Performance on large dataset	Slow	Fast	Fast

10.7 APPLICATIONS OF ANN

Now that we have an overall understanding of the functioning of ANN, let us look at its applications across different sectors in the industry.

10.7.1 IMAGE PROCESSING AND CHARACTER RECOGNITION

A very crucial application of ANN is in the area of image processing. Researchers have used neural network for detecting breast cancer [6] in its early stage, by taking image obtained from mammography as input. The image segmentation is applied before feeding it a 3-layered ANN to detect region where malignant and benign lesions is seen. Another research involves applying Gabor wavelet to the mammographic image for detecting cancer [7]. The researchers claim that transforming the images using Gabor wavelets increases the accuracy of prediction to 97% on the unseen data images. Self-organizing maps (SOM) used in combination with NN for detecting breast cancer being benign and malignant [8]. Another approach to identify malignant nodules is through pre-processing images in order to extract five morphological features depicting shape, edges, and darkness of the nodules. Following this, images are subjected to classification via ANN post k-fold cross-validation [9, 10]. Another popular approach involved using CNN with K means clustering to detect breast cancer nodules in mammographic images. The hybrid approach was found to be providing better results [11]. It is widely used in the detection of tumors in humans, thus helping the doctors to make quicker and efficient decision based on the location and size of the tumor.

It is also used in facial detection, which can be used to identify a criminal in hiding or even on Facebook to tag your friends. They are also used in processing images from satellites either for defense purposes or in agricultural fields to identify whether a plant is healthy or not. Thai Hoang Le [12] utilized a novel approach by using AdaBoost and ANN for detecting faces in image. Once faces are detected, he performs face alignment by creating a three-layered MLP model equipped with linear function across each layer. Further, in order to extract facial features like eyes, nose, mouth, etc., in an image, the author explored the usage of ICA. The author has reported a fairly decent accuracy in his research to solve face recognition challenges.

To deal with these kinds of challenges, different CNN architectures are also used since these architectures are more robust in dealing with image data. Handwritten character recognition is mainly used to verify if a signature in a document is legitimate or not. Researchers have developed a writer-identification system using deep CNN by building a pipeline for data acquisition, data processing, and classification [13]. Likewise, these models can also be used to identify the number plates of a vehicle.

10.7.2 STOCK PRICE PREDICTION

Predicting and investing in stock exchange has a prominent presence in the business side of the world. They would expect good profit margins for their investments, but manually predicting has a higher risk of losing money since there are lots of variations happening consistently due to various external factors. Thus an application is needed to monitor these changes, understand the underlying patterns, and come up with accurate predictions on the future trend of the market. Thus LSTM, a variation of RNNs, which deals with sequential data are used. Sophisticated algorithms with respect to this are built in the industries, which aid them in investing in stocks wisely. Feedforward NN and recurrent NN are explored by Trippi and Turban [26], and a similar effort was given by Walczak [27] and by Shadbolt and Taylor [28] on empirical foreign exchange rate for matters on financial forecasting. Mingyue Qiu and Yu Song [14] contributed a powerful hybrid model using genetic algorithm and ANN to predict the movement of a stock market index. In 2017, Mehdi and Zahra [29] build a two series hybrid model using ARIMA and ANN to develop time series forecasting on financial data [15].

10.7.3 LOAN APPLICATION EVALUATION

A very interesting application of neural network is witnessed by a group of researchers where the creditworthiness of loan applicants are determined by training a NN. Banks provide loan to the users based on various factors like the applicant's characteristics such as his age, account type, income, nationality, residency, companies' type, guarantor, job experience, and the DBR (i.e., debt balance ratio that measures the applicant's repaying ability). When loan applicants are more in number, it becomes tedious and time-consuming for a normal human being. Also, banks would like to minimize the failure rate of loan applications and maximize the returns on the loan issued. Thus to speed up the process, Neural Networks are used to classify whether an applicant is eligible for a loan or not. Shorouq and Saad [16] from the University of Jordan used simple NN to assist Banks in Jordan for scrutinizing loan applications and they witness fairly decent accuracy from their model.

10.7.4 HEALTHCARE

Health care organizations in developed countries are facing challenges related to the delivery of a value-based, patient-centric facilities to patients in and around the world. A study carried out by Bartosch-Härlid et al. [17] investigated that there were more than 11 scientific articles that are published between 1993 and 2003 that demonstrated the best usage of ANNs on pancreatic diseases. ANN was explored for prediction and scientific analysis in domains of cardiovascular, telemedicine, and organizational behavior. Further, it was observed that ANN was predominantly used in the area of cardiovascular medicine: for analysis and remedy of coronary artery disease, standard interpretation of electrocardiography, cardiac image evaluation and cardiovascular drug dosing [18]. Telemedicine gives fitness care solutions for patients in remote areas by building health monitoring devices that are designed to prevent, diagnose, manage ailment [19] and also include smart devices that are equipped with ML techniques to predict medical conditions like high blood pressure [20]. Preliminary analysis of high-risk patients (for sickness or attributes) the use of neural networks offer clinic administrators with a cost-effective tool in time and useful resource management [21].

Image data in the medical field is on the rise thanks to X-rays, MRI, CAT scans, etc. Going through these images manually can be time consuming and require a high level of expertise in doctors to analyze and draw conclusions from these images. This is where Neural Networks come into the picture. CNNs, are widely used to deal with these images, thus helping doctors identify patterns and symptoms, prediction of disease, which leads to faster clinical decisions, hence increasing the chances of survival while reducing the treatment cost. These models are widely used in the detection of tumor and predict whether they are benign or malignant.

10.7.5 AGRICULTURE

Agriculture is one more important sector in which neural networks is making its presence felt. They are implemented right from species breeding and selection, where a neural network model is built to analyze decades of field data and try to predict which genes contribute to productivity. It is also used in weed detection which uses CV to identify and segregate between a productive plant and unwanted plants. Neural networks are also used to predict crop yield based on various external and internal factors. They are also used to identify and predict disease from which the crop is suffering, thus immediate action can be undertaken to control it.

10.7.6 SELF DRIVING CARS

Self-driving cars is currently one of the hot topics of research in the automobile industry. These cars are driven autonomously with little to no human intervention thanks to the implemented ML algorithms that have been trained in real-world scenarios such as 'stopping when the signal light is red' or detecting other vehicles or other obstacles around its proximity, by collecting the information via sensors and cameras attached onto the surface of the car in real-time. Some of the major firms which are leading the way in this research are Tesla, Google cars, Bosch, etc. As far as the implementation is concerned, multiple Deep Neural Networks or DNNs are used, each designed to perform specific tasks, thus chances of road hazard are minimal. Some of the DNNs used by NVIDIA in their systems are as follows:

- **PathFinders:** This section contains DNNs which help in determining and planning the next path ahead.
- **OpenRoadNet:** This DNN measures and identifies the drivable space of other vehicles which is on the same lane as well as the neighboring lane.
- **PathNet:** This DNN is designed to highlight the drivable path of the vehicle, even in the absence of the lane markers.
- **LaneNet:** This DNN is designed to detect lane lines and other markers which determine the car's path.
- **MapNet:** This DNN is designed to lanes and landmarks in order to create and update high-definition maps.
- **Object Detection and Classification:** This section contains DNNs which help in determining obstacles and road signs (Figure 10.12).
- **DriveNet:** This DNN is implemented to perceive other cars on the road, pedestrians, traffic lights and signals, etc. They do not read the color of the light or the type of sign.
- **LightNet:** This DNN is designed to determine the state of the traffic light, i.e., to determine whether they are red, yellow or green.
- **SignNet:** This DNN is implemented to identify the type of road sign.
- **WaitNet:** This DNN is designed to decide whether the whether the vehicle has to stop and wait, example being when it is an intersection.

Thus, these are some of the functionalities involved in the making of self-driving cars. Although they are not completely into the commercial phase, the rapid advancements in research and technology will ensure this will be a major breakthrough in the market in the upcoming years.

FIGURE 10.12 Image of a self-driving car identifying various objects on its path.

10.7.7 CLASSIFICATION OF ASTRONOMICAL OBJECTS

Astronomical objects are naturally occurring structures observed in the universe. These comprises of stars, galaxies, quasars, planets, moons, asteroids, comets, etc. Various projects are launched over the period of time to conduct sky surveys for collecting photometric observations of astronomical bodies with the help of high-resolution telescopes. The telescopes are equipped with special filters to capture electromagnetic radiation emitted from celestial objects in the sky. These telescopes have captured a spectra of stars, galaxies, and quasars, and their catalogues are released in the public domain to carry out research on solving different astronomical challenges. The most recent research conducted on SDSS sky survey data involved researchers to use CNN to estimate photometric redshift and their probability distribution function [22, 23]. Prior to this, k-NN, and random forest were also explored for such problems. A CNN architecture is built by taking batches of images of different bands (u – ultraviolet, g – gamma, r – radio, i – infrared, z – z band) are the five SDSS filters used to obtain these images). Convolution and pooling layers are built applying a 5×5 filter on image and PReLU (parametric ReLU) and ReLU are used as activation functions. Softmax layer is added as the last layer and cross-entropy loss function is used while training the network. As indicated by the researchers, PDF's of redshift obtained from CNN model are observed to be fairly reasonable values.

Another set of experiment was performed using the SBAF activation function [24] for classification of exoplanets into habitable and non-habitable classes. SBAF used on a three-layered MLP (multi-layered perceptron) has demonstrated a practically distinct accuracy based on the features fed into the neural network.

KEYWORDS

- applications of neural networks
- artificial neural networks
- gradient descent
- multilayer perceptron
- neural networks
- neuron

REFERENCES

1. Natalia, A. G., Djai, B. H., René, W., & Matthijs, B. V., (2018). *Large and Fast Human Pyramidal Neurons Associated with Intelligence*. eLIFE.
2. Frank, R., (1960). Perceptron simulation experiments. *Proceedings of the IRE, 48*(3).
3. Yann, L., Patrick, H., Léon, B., & Yoshua, B., (1999). *Object Recognition with Gradient-Based Learning*. Shape, contour and grouping in computer vision, Springer Publications.
4. Sepp, H., & Jürgen, S., (1997). Long short-term memory. *Neural Computation* (Vol. 9, No. 8). The MIT Press Journals.
5. Rumelhart, D. E., Hinton, G. E., & Williams, R. J., (1985). *Learning Internal Representations by Error Propagation*. MIT Press.
6. Mehdy, M. M., Ng, P. Y., Shair, E. F., Saleh, N. I. M., & Gomes, C., (2017). Artificial neural networks in image processing for early detection of breast cancer. *Computational and Mathematical Methods in Medicine, 2017*, 15. Article ID: 2610628. https://doi.org/10.1155/2017/2610628.
7. Lashkari, (2010). Full automatic microcalcification detection in mammogram images using artificial neural network and Gabor wavelets. In: *Proceedings of the 6th Iranian Conference on Machine Vision and Image Processing (MVIP '10)*. Isfahan, Iran.
8. Dar-Ren, C., & Chang, R., & Huang, Y. L., (2000). Breast cancer diagnosis using self-organizing map for sonography. *Ultrasound Medicine Biology, 26*(3), 405–411. 10.1016/S0301-5629(99)00156-8.
9. Joo, S., Yang, Y. S., Moon, W. K., & Kim, H. C., (2004). Computer-aided diagnosis of solid breast nodules: Use of an artificial neural network based on multiple sonographic features. *IEEE Transactions on Medical Imaging, 23*(10), 1292–1300.
10. Joo, S., Moon, W. K., & Kim, H. C., (2004). Computer-aided diagnosis of solid breast nodules on ultrasound with digital image processing and artificial neural network. In: *Proceedings of the 26th Annual International Conference of the IEEE Engineering in Medicine and Biology Society (EMBC '04)*, (Vol. 2, pp. 1397–1400). IEEE, San Francisco, Calif, USA.
11. Zheng, K., Wang, T. F., Lin, J. L., & Li, D. Y., (2007). Recognition of breast ultrasound images using a hybrid method. In: *Proceedings of the IEEE/ICME International Conference on Complex Medical Engineering (CME '07)* (pp. 640–643). IEEE, Beijing, China.
12. Thai, H. L., (2011). Applying artificial neural networks for face recognition. *Advances in Artificial Neural Systems, 2011*, 16. Article ID: 673016, 16 pages. https://doi.org/10.1155/2011/673016.
13. Weixin, Y., Lianwen, J., & Manfei, L., (2015). Chinese character-level writer identification using path signature feature, drop stroke and deep CNN. In: *13th International Conference on Document Analysis and Recognition (ICDAR)*. doi: 10.1109/ICDAR.2015.7333821.
14. Qiu, M., & Song, Y., (2016). Predicting the direction of stock market index movement using an optimized artificial neural network model. *PLoS One, 11*(5), e0155133. doi: 10.1371/journal.pone.0155133.

15. Khashei, M., & Hajirahimi, Z., (2017). Performance evaluation of series and parallel strategies for financial time series forecasting. *Financ. Innov., 3*, 24. doi: 10.1186/s40854-017-0074-9.
16. Shorouq, F. E., & Saad, G. Y., (2010). Applying neural networks for loan decisions in the Jordanian commercial banking system. *IJCSNS International Journal of Computer Science and Network Security, 10*(1).
17. Bartosch-Härlid, Andersson, B., Aho, U., Nilsson, J., & Andersson, R., (2008). *Artificial Neural Networks in Pancreatic Disease.* BJS Society. https://doi.org/10.1002/bjs.6239.
18. Itchhaporia, D., Snow, P. B., Almassy, R. J., & Oetgen, W. J., (1996). Artificial neural networks: Current status in cardiovascular medicine. *J. Am. Coll. Cardiol., 28*(2), 515–521.
19. Lymberis, A., (2003). Smart wearables for remote health monitoring, from prevention to rehabilitation: Current R&D, future challenges. In: *4th Annual IEEE Conference on Information Technology Applications in Biomedicine.* United Kingdom.
20. Kwong, E. W., Wu, H., & Pang, G. K., (2016). A prediction model of blood pressure for telemedicine. *Health Informatics J.* pmid:27496863. https://doi.org/10.1177/1460458216663025.
21. Nolting, J., (2006). Developing a neural network model for health care. *AMIA Annu. Symp. Proc.,* 1049.
22. Johanna, P., Bertin, E., Treyer, M., Arnouts, S., & Fouchez, D., (2019). Photometric redshifts from SDSS images using a convolutional neural network. *Astronomy and Astrophysics, 621.* https://doi.org/10.1051/0004-6361/201833617.
23. Snehanshu, S., Archana, M., Kakoli, B., Surbhi, A., & Suryoday, B., (2008). A new activation function for artificial neural net-based habitability classification. *Proceedings in International Conference on Advances in Computing, Communications, and Informatics (ICACCI),* 1781–1786.
24. Snehanshu, S., Nithin, N., Archana, M., & Rahul, Y., (2019). *Evolution of Novel Activation Functions in Neural Network Training with Applications to Classification of Exoplanets.* arXiv preprint arXiv:1906.01975.
25. Box, G. E. P., & Hunter, J. S., (1957). Multi-factor experimental designs for exploring response surfaces. *Ann. Math. Statist., 28*, 195–241.
26. Trippi, R. R., & Turban, E. (Eds.). (1992). *Neural Networks in Finance and Investing: Using Artificial Intelligence to Improve Real World Performance.* McGraw-Hill, Inc.
27. Walczak, S. (2001). An empirical analysis of data requirements for financial forecasting with neural networks. *Journal of Management Information Systems, 17*(4), 203–222.
28. Taylor, J. G. (2002). Neural Networks. *In Neural Networks and the Financial Markets* (pp. 87–93). Springer, London.
29. Khashei, M., & Hajirahimi, Z., (2017). Performance evaluation of series and parallel strategies for financial time series forecasting. *Financ. Innov., 3*, 24. doi: 10.1186/s40854-017-0074-9.

CHAPTER 11

COMPARING WORD EMBEDDINGS ON AUTHORSHIP IDENTIFICATION

TARUN KUMAR DUGAR, S. GOWTHAM, and
UDIT KR. CHAKRABORTY

Department of Computer Science and Engineering, Sikkim Manipal Institute of Technology, Sikkim Manipal University, Sikkim, India, E-mail: udit.c@smit.smu.edu.in (U. Kr. Chakraborty)

ABSTRACT

Authorship identification is a task which has found historical importance and several widespread applications across different fields. It basically functions as a technique to help attribute the ownership of an anonymous piece of text. Embeddings are discrete fixed-length representations of texts which aim to capture maximum information within these numeric representations. The embedding techniques used in this chapter include Doc2Vec, GloVE, and FastText. The primary focus of the work is to estimate the performance of the word embedding techniques for the task of authorship identification. Deep and artificial neural networks (ANNs) are used to help classify these authors on the basis of these embeddings and their respective results have been reported. The experimental results compared among the competing techniques gives fruitful insight into the performance of the embeddings as also the task of authorship identification.

11.1 INTRODUCTION

The task of associating a piece of text, based on the attributes of their style of writing, to its original author can be referred to as 'authorship identification.' These attributes, which help identify the author, can help distinguish the authors' style, and they become especially useful in a context where the author of the text is not explicitly discernible. Several methods and techniques have been employed in the past to garner such useful attributes. Methods ranging from using Latent Dirichlet Association to identify the texts [1] quantitative information like stylometric variables [2] or statistical evaluation [3] to using qualitative information [4] have been applied to achieve this task. These techniques have proved useful time and again in tasks such as profiling authors, forensic analysis, applications for intelligence, plagiarism detection, to name a few. Recent advances have observed this field exploit a wide range of areas like machine learning (ML), natural language processing, information retrieval systems, etc. Chen Qian et al., in their chapter on deep learning-based authorship identification [5] reported a 69.1% and 89.2% accuracy on two different datasets. This chapter intends to realize the performance of deep artificial neural networks (ANNs) while using text embedding techniques.

In this text, the authors exploit the abilities of deep ANN to help identify the authors from the extracted set of features of the text. ANNs are represented by a set of interconnected nodes, called neurons, which are vaguely modeled after neurons in a brain. These neurons are linked to each other through connections called edges. Each edge has an adjustable weight associated with it and these weights control the strength of the signal that passes through that edge. Every bunch of neurons is modeled into layers which are interconnected through edges. ANN's are especially useful in their form of data processing which allows them to extract characteristics or feature sets from the data that it processes.

ANNs accept inputs only as numbers. To meet this requirement, all data has to be suitably converted to some numerical representation. Embeddings are methods or techniques of converting non-numerical data to numerical representations, having all features of the input embedded in the output. Neural Embeddings take discrete representations such as texts and output fixed-length continuous vector representations for the same. Recent advances in this field have seen the emergence of several techniques and architectures to obtain neural embeddings. Various approaches have been employed to achieve efficient embedding for neural networks and models using context-based, character-based, word analogies [6], or even statistical information exist.

Each technique has had its share of success and their spectrum of application. The choice of the ideal embedding for a given task, however, is still largely corpus-oriented and problem-centric.

The work reported in the chapter compares the performance of three different embedding techniques namely Doc2Vec, GloVE, and FastText and using Stylometric features, with an intent to identify the best for the task.

11.2 LITERATURE SURVEY

Authorship Identification has historically found wide applications in fields like humanities, linguistics, forensic studies, etc. Recent advances in Authorship Identification have seen a wide range of methods being applied to achieve improved results. Techniques ranging from using textual data generated from the texts to using advanced ML techniques have been employed to perform this task. Smita Nirkhi et al. used generalized features from previous research works to extract information and used Support Vector Machines [7] for classification. Ilker Nadi Bozkurt et al. performed Authorship Identification on Turkish newspaper Milliyet using both supervised and unsupervised learning techniques with various feature extraction methods and recorded the observation [8]. Ahmed M. Mohsen et al. used a Stacked Denoising Auto-Encoder to extract the features from the text and a Support Vector Machine to classify them accordingly [9]. Zhou and Whang [10] used recurrent neural networks (RNNs) at the sentence level to capture useful features from a piece of text and were able to obtain an approximate F_1 score of 0.6 with word dimensions of 300.

With the emergence of new and improved embedding techniques, neural embeddings can now capture more semantic information within these vector representations. Doc2Vec [11], an embedding technique proposed by Quoc Le and Tomas Mikolov, can generate state-of-the-art vectors for documents using a prediction-based model. FastText [12] is an embedding technique created at Facebook AI Research, which leverages sub-word information, i.e., it breaks down a word into character n-grams and each word may be represented as a sum of such discrete n-grams. Unlike the aforementioned techniques, GloVE [13] is another embedding technique which tends to leverage statistical information to create word vectors emphasizing on creating a vector space with a meaningful substructure. Hemayet Ahmed Chowdhury et al. [14] performed all the aforementioned word embeddings on Bengali Literature and used neural networks like convolutional and RNN for classification of Authors. In this chapter, we aim to compare and contrast the techniques like stylometry and word embeddings to help perform Authorship Identification.

11.3 ABOUT EMBEDDINGS

Word embeddings are techniques of representing words in vector space. These may be applied on individual words, sentences, paragraphs or even complete documents. Embeddings find special application in ANNs where only numerical values are accepted as inputs. Since embeddings converts words or other text to vectors, the vector embedding gives a suitable alternative for words. The efficiency of an embedding technique, however, is measured through its ability to map words with similar meaning or having higher co-occurrence closer to each other in the vector space.

11.3.1 DOC2VEC

Doc2Vec is an unsupervised technique which was developed by Quoc Le and Tomas in 2013 [11]. It is used to generate fixed length vector representations for documents of variable sizes. This embedding technique can to a great extent, capture the semantic relations between words, i.e., if mapped into vector space, words with similar semantic meaning are mapped closer to one another. The vectors are generated by a neural network which uses stochastic gradient descent. In this technique, the gradient is obtained by using back-propagation and can be used both for words and paragraphs. Using doc2vec, it is also possible to learn a linear matrix to translate words/phrases among languages.

In this method, each paragraph is given a unique vector representation, whereas the word representations are shared among the documents. This is where the technique heavily relies on a word embedding technique known as Word2Vec [12]. Doc2Vec uses Word2Vec to obtain fixed-length representations of the words which in turn help generate the same at a document level. The paragraph vector is obtained from these word vectors, which are then trained on the model until it converges. It uses two models to generate results-distributed memory and distributed bag of words. In the

Distributed Memory model, a word is randomly sampled from a sentence and the model tries to predict the word given the context. Whereas in the Distributed Bag Of Words model, for each iteration, we randomly sample words from a sentence and form a classification task where we ignore the context of the words but force the model to predict them.

Usually, the Distributed Memory model can achieve state-of-the-art results by itself and works better than the Distributed Bag of Words. But it is recommended to use a combination of the two models to achieve the best results. Also, unlike the DM model, which gives equal importance to words that occur infrequently, the DBOW creates a classification task among words (Figure 11.1).

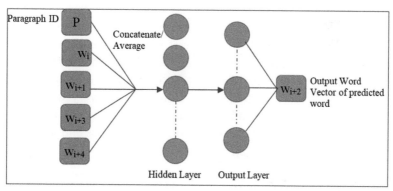

FIGURE 11.1 Doc2Vec model (distributed memory).

11.3.2 FASTTEXT

FastText [13] is an embedding technique developed by Piotr Bojanowski, Edouard Grave, Armand Joulin, and Tomas Mikolov at the Facebook AI Research. The basic idea behind FastText relies on the way that it extracts the information from the words. FastText, unlike prediction or statistic-based methods takes a different approach from other embedding techniques. While other embedding techniques ignore the internal structure of the words, which is very important in languages where the same word may have several different forms, FastText provides improved vector representations by leveraging character-level information.

In FastText, each word is represented as a set of character n-grams, i.e., each word is broken down into parts of n consecutive characters which serve as the basis for evaluation. Considering the word "broth" for a value of $n = 3$, the char n-grams may be given as follows: <br, bro, rot, oth, th> and <broth>, where < and > represent extremes of the word. The word itself is also included in the set of char n-grams to obtain its representation. The value of n, i.e., the size of the n-grams must be chosen carefully as it is a deciding factor in the performance of the model.

The advantage of using FastText lies in its handling of rare or infrequent words. Unlike other techniques, each word has not already been assigned a distinct vector, and thus, better representations can be generated for words which are out of vocabulary for a corpus.

Another advantage is that since it calculates n-grams, FastText tends to capture the internal structure of the words. Since some languages have several forms of the same word, FastText provides improved representational power for such words because it deals with character level information.

A drawback is that all the input must be pre-processed with the help of a parser before it may be fed to train the model. Another problem that might come up while using FastText is the amount of memory it uses up as it needs to store vectors for each n-gram for every word. Despite the intensive computation and large memory requirements, FastText can perform better than several other embedding techniques and provide state of the art results (Figure 11.2).

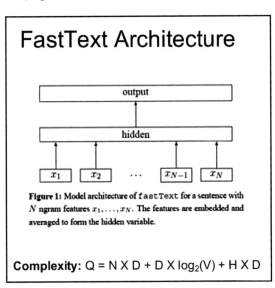

FIGURE 11.2 Architecture of FastText model.

11.3.3 GLOVE

Global Vector Representation or GloVE [14] is a statistical neural embedding generation technique which relies on word probability count data from a corpus. It was developed by Jeffery Pennington, Richard Socher, and Christopher D. Manning at the Stanford University. Global vectors (GloVe) is another method to produce embeddings for words in a document. They mitigate the drawback of Word2Vec that is they do not use Global Statistical information of the words in the given text samples yet they produce Dimensions of Meaning of the tokenized words from the given text samples.

Many models have been created which leverage the word count statistics to perform unsupervised learning of word representations, each having its own unique way of the way they make use of the data. Glove makes use of ratios of probabilities of co-occurrence of each word. Let "*i*" and "*j*" be two words. If a given word is similar in meaning to "*i*" but not "*j*," the ratio for that word will be higher (and vice versa). For words related to both "*i*" and "*j,*" the ratio should be close to one. It relies on the fact that "compared to raw probabilities, the ratio is better able to distinguish between relevant and irrelevant words" (Figure 11.3) [8].

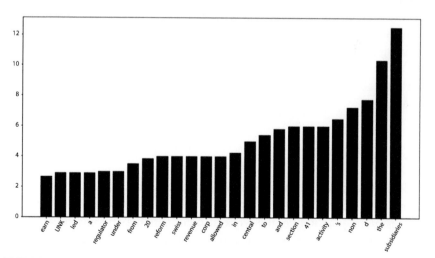

FIGURE 11.3 Graph depicting the words synonymous with the word "bank."

GloVE outperforms models like Word2vec as it uses statistical information, i.e., word count data, and also manages to keep a meaningful linear substructure of the words. One of the cons of this model could be the fact that it has a larger memory footprint when compared to models like word2vec. Irrespective of the speed of the operation, GloVE can consistently provide better results when compared to other techniques. These two methods, i.e., statistic-based and prediction based both basically rely on the same idea based on word co-occurrences, but the count-based method used by glove is known to deliver consistent results across varying types of tasks.

11.4 DATASET

The dataset used for the study was the Reuters_50_50 dataset, which is a subset of the Reuters Corpus Volume 1. An archive created and made available for research purposes by Reuters [15], the RCV I consists of over 800,000 manually categorized newswire stories. It consists of 50 articles each of 50 different authors and has been used extensively as a standard dataset for authorship identification research [16]. The articles were carefully chosen, keeping the factor of topic fairly common thus minimizing its effect on the dataset. The dataset has been divided into two non-overlapping sets of articles, namely testing and the training sets. The training dataset had 80 articles, while 20 articles per author was earmarked for testing. The choice of the dataset was based on the fact that the entire set was labeled with at least one subtopic of the class CCAT (corporate/industrial).

11.5 APPROACH

This section describes the approach taken to solve the problem at hand using each of the following methods. Apart from the embedding techniques mentioned above, the work also contained a stylometric approach to measure the efficacy of the neural embeddings against a more traditional method. The basic aim was to test the efficacy of each style in identifying authorship of articles. To that end, experiments were conducted using ANNs choosing the most accurate results from the host of experimental readings. The following section discusses in detail each of the methods.

11.5.1 STYLOMETRIC

Defined by the Oxford dictionary as, the statistical analysis of variations in literary style between one writer or genre and another, stylometry primarily is the study of the style of writing. However, stylometric analysis has found successful applications in music and painting as well.

The science of stylometry is based on the observation that individuals have a distinct style of writing which is exposed through parameters like word usage, punctuation, sentence lengths, among others, and can be used to identify the authorship of any given text.

The stylometric approach makes use of textual variables to obtain the features from the author's articles. The choice of variables hence becomes very crucial to the performance of the task. The following textual variables have been chosen to help capture the author's style of writing.

- Number of occurrences of topic in the text.
- Number of occurrences of "I."
- Number of occurrences of "We."
- Number of commas in the text.
- Number of full stops in the text.
- Number of exclamation marks in the text.
- Number of quotations in the text.
- Average lengths of sentences.
- Length of the article (measured in words).

These parameters served as the basis for the numeric representation of the text. Feedforward ANNs were used to help classify the authors with the generated vectors acting as input to the network. This approach proved to be particularly insightful as to the extent to which a discrete set of textual variables can represent the author's style of writing.

Extensive experiments were carried out with stylometric data on different model neural nets. The results have been tabulated in Table 11.1. The parenthesized annotations stand for r-ReLu, si-Sigmoid, and so-Softmax. The highest accuracy level was found with a network of 64, 32, and 50 nodes with ReLu, ReLU, and Softmax activations, respectively. The observations do not bring out any correlation between the number of epochs and accuracy. However, it may be noted that the accuracy of stylometric approaches depend largely on the choice of parameters which needs to be chosen after careful study of the text.

TABLE 11.1 Experimental Results with Stylometric Parameters

Hidden Layers	Nodes in Layers	Epochs	Accuracy
2	64(r) + 32(r) + 50(so)	500	15.4
2	64(r) + 32(r) + 50(so)	1000	16.2
2	128(r) + 32(r) + 50(so)	1000	14.000
2	128(r) + 32(r) + 50(so)	5000	15.6
2	128(r) + 64(r) + 50(so)	500	13.600
2	128(r) + 64(r) + 50(so)	1000	15.2999
2	128(r) + 64(r) + 50(so)	5000	2.0
3	128(r) + 64(r) + 32(r) + 50(so)	500	14.499
3	128(r) + 64(r) + 32(r) + 50(so)	1000	13.0
3	128(r) + 64(r) + 32(r) + 50(so)	5000	14.899
3	128(r) + 64(r) + 64(r) + 50(so)	500	14.49
3	128(r) + 64(r) + 64(r) + 50(so)	1000	12.6
4	128(r) + 64(r) + 64(r) + 32(r) + 50(so)	500	13.0
4	128(r) + 64(r) + 64(r) + 32(r) + 50(so)	1000	13.60
2	128(si) + 32(r) + 50(s)	1000	14.39
2	128(si) + 32(r) + 50(s)	5000	15.1
2	128(si) + 64(r) + 50(s)	500	15.1
2	128(si) + 64(r) + 50(s)	1000	13.200

11.5.2 DOC2VEC

Doc2Vec is a word embedding generation technique which helps generate fixed-length vectors for documents. It is a technique derived from Word2Vec, a word-vector generation method. The idea behind Doc2Vec is that apart from the word vectors, another vector termed paragraph-id is added. This paragraph id marks the serial of the paragraph from which the selected portion of text has been taken. The vectors are generated by using a distributed bag of words model or a distributed memory model. A hidden layer with a Softmax function is also employed and whenever a new document is submitted and fixed weights were used to compute the vector.

In a nutshell, the objective of Doc2Vec can be stated as:

$$\frac{1}{T}\sum_{t=k}^{T-k} \log p\left(\omega_t \mid \omega_{t-k}, \cdots \omega_{t-k}\right) \tag{1}$$

A detailed study on the Doc2Vec embedding conducted over the same dataset has been published by the authors [17]. The results have been compared here.

11.5.3 GLOVE

Glove, owing to its statistical approach, tends to generate a much-nuanced representation for the words in the corpus. 75% of the corpus was used to serve as training data, to help learn the features while the rest 25% was used for testing. A word-vector map is then generated for each author. From each of these word vectors, we pick the co-ordinate wise minimal and maximal vectors and append them. These vectors serve as the document representation and hence serve as input to the ANNs. Apart from this, the Stanford NLP group have also provided with a few more models, which were trained on millions of words from large corpora of data. These helped serve as additional avenues for exploration for the task at hand.

GloVe is a count-based log-bilinear model with a weighted least-squares objective. It takes into account Global Statistical Information by constructing a co-occurrence matrix. The matrix is constructed by taking into account the frequency of the words that occur together in the given text sample for a window size of 4. The window size can be changed. Here, a value of 4 is taken considering the length of the text sample. Table 11.2 shows a sample of the co-occurrence matrix. The first row depicts the column name.

TABLE 11.2 Sample Co-Occurrence Matrix

8	9	10	11	12
17.33	10.25	20.75	11.83	15.33
19.33	13.41	7.25	19.41	13.58
34.08	16.08	11.41	6	5.83
3.416	71.16	9.5	4.5	19.91
68.91	0.33	5.25	5.66	3

Understanding the meanings of a word makes GloVe embedding works better with larger datasets and even when the numbers of epochs are also lower. This is evident from Figure 11.4, which shows that for a smaller dataset that was used and with fewer epochs of about 250 it gives an accuracy of 78%. Although they have higher fluctuations of loss values, they still end up getting the highest accuracy. The performance measure with a larger dataset is as shown in Figure 11.5. The same dataset with an epoch of 500, there is a steady increase in the loss value after the epoch 400 and ends up with an accuracy of 63%.

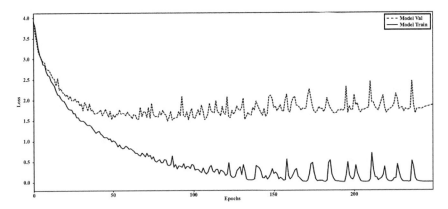

FIGURE 11.4 GloVe for 250 epochs.

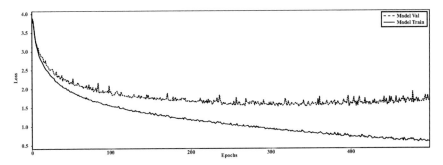

FIGURE 11.5 GloVe for 500 epochs.

Comparing Word Embeddings on Authorship Identification

Table 11.3 lists the accuracy values for the trial runs for GloVe having embedding size 100 using the Skip-Gram model. The alphabets within parentheses in the second column refer to the activation function used. Here (r) stands for ReLU and (s) for Softmax functions. The numbers of nodes have been varied and observations tabulated. It may be noted that for the same configuration, a higher number of epochs actually results in a lower accuracy value, which may be attributed to overfitting.

TABLE 11.3 Accuracy Values for Trial Runs of GloVe

Number of Hidden Layers	Number of Nodes in Layers	Epochs	Validation Accuracy
3	64(r) + 32(r) + 16(r) + 50(s)	100	46.91
3	64(r) + 32(r) + 16(r) + 50(s)	250	57.46
3	64(r) + 32(r) + 16(r) + 50(s)	500	63.43
5	512(r) + 256(r) + 128(r) + 64(r) + 32(r) + 50(s)	250	78.64
5	512(r) + 256(r) + 128(r) + 64(r) + 32(r) + 50(s)	500	67.81
7	1024(r) + 512(r) + 512(r) + 256(r) + 128(r) + 64(r) + 32(r) + 50(s)	500	66.51
7	2048(r) + 1024(r) + 512(r) + 256(r) + 128(r) + 64(r) + 32(r) + 50(s)	250	74.95

Table 11.4 shows the confusion matrix for two authors A_1 (Aaron Pressman) and A_2 (Alan Crosby). The vector for A_1 created using GloVe returns 100% accuracy for A_1 and 47.99% for A_2. Similar observations were made for A_2 as well. Table 11.5 shows the confusion matrix for all four authors. While embeddings of individual authors return 100% accuracy with their own data, the matrix shows the average of other authors for a given author vector.

TABLE 11.4 Confusion Matrix for Two Authors

	A_1 Data	A_2 Data
A_1 Vector	100	47.99
A_2 Vector	73.0	100

TABLE 11.5 Confusion Matrix for Four Authors

	$A_2A_3A_4$	$A_3A_4A_1$	$A_4A_1A_2$	$A_1A_2A_3$
A_1	23.00	–	–	–
A_2	–	32.499	–	–
A_3	–	–	18.500	–
A_4	–	–	–	23.499

11.5.4 FASTTEXT

FastText essentially can train models using a continuous bag of words or a Skip-gram model using different kinds of loss functions. It tends to scale well with huge corpora of data owing to methods like multithreading for taking data input, automatic pruning, a flag-based pruning and sampling tables. But despite these features, factors such as the character n-gram size to be chosen, the minimum and maximum vocabulary size flags, the default threshold, etc., do play a major role in the quality of the vector turnout. A sampling table is also created to discard frequent words, as they tend to offer less information when compared to rare words. The probability that a word can be included in this sampling table, i.e., get discarded is based on a function. Selecting appropriate values for the constants and thresholds for such functions also plays an important role in fine-tuning the vectors. What is important to note is the discarding of words only occurs in an unsupervised model. With thorough experimentation with the aforementioned features, we train both a skip-gram and a continuous bag of words model with varying vector sizes to achieve the best possible results. These vectors are then used as input to the deep neural network to help classify the authors appropriately.

Different combinations of vector lengths and variations of the embedding were experimented upon. There being no standard rule of deciding upon the final structure of the ANN, a number of different structured were also tried. Results were tabulated incrementally changing the number of nodes in different layers number of layers similarly altered across models. Table 11.6 presents a summary of observations across a sizeable number of alternate layouts.

TABLE 11.6 Summary of Observations for FastText

Model	Embedding Size	No. of Hidden Layers	Number of Nodes in Layers	Epochs	Validation Accuracy
CBOW	40	2	64(r) + 32(si) + 50(s)	100	87.73
Skip-gram	40	2	64(r) + 32(r) + 50(s)	100	80.49
CBOW	60	2	64(r) + 32(r) + 50(s)	100	85.61
Skip-gram	60	2	64(r) + 32(r) + 50(s)	100	82.52
CBOW	100	2	512(r) + 256(r) + 50(s)	100	91.79
Skip-gram	100	2	64(r) + 32(r) + 50(s)	100	85.93

Among the two variants of FastText, experimental results reveal that the common bag of words (CBOW) approach outperforms the Skip-Grams approach for all embedding sizes. It can also be seen that the validation accuracy improves with the size of the embedding vector. This may be attributed to the fact that a longer embedding stores more information about the text than would a shorter more compact one.

11.6 RESULTS AND DISCUSSIONS

Individual styles and embeddings having been implemented and results sorted, the best performing machines and approaches had been tabled. Table 11.7, lists the best performing architectures across different approaches taken. The parameters considered for comparison are accuracy, number of hidden layers, number of epochs, and number of hidden nodes.

TABLE 11.7 Best Performing Architectures Across Approaches

Approach	Accuracy	No. of Hidden Layers	Epochs	No. of Hidden Nodes
Stylometric	16.2	2	1000	96
GloVe	78.64	5	250	992
FastText-CBOW 40	87.73	2	100	96
FastText-CBOW 60	85.61	2	100	96

TABLE 11.7 *(Continued)*

Approach	Accuracy	No. of Hidden Layers	Epochs	No. of Hidden Nodes
FastText-CBOW 100	91.79	2	100	768
FastText-SkipGram 40	80.49	2	100	96
FastText-SkipGram 60	82.52	2	100	96
FastText-SkipGram 100	85.93	2	100	96
Doc2Vec	67.6	2	2500	192

Apparently, from the data presented, it seems that FastText CBOW with vector size 100 is the best as it delivers the most accurate results at 91.79%. However, a closer study reveals a few more interesting facts. Leaving aside the Stylometric and Doc2Vec approaches, the remaining approaches return accuracies in the range of 79% to 91%, which for a task of magnitude and complexity as considered, can be considered to be an allowable margin. However, GloVe appears to be more resource-hungry compared to any other approach and utilizes close to a 1000 nodes spread over five layers. All others with comparable accuracies use only two layers. The best performing approach, FastText CBOW 100 uses 768 nodes over two layers in a 100 epochs.

The experimental results, however, do not show inconclusively, the supremacy of any of the given approaches. While one may argue at the competence of the stylometric method, it remains to be understood how exactly the parameters influence the performance. One may even be intrigued by the subtle choices made by the expert human observer in choosing the optimal parameters to render the best outcome. All others, including the apparent outlier Doc2Vec, are close enough to be contending for selection for a given task. It is clear that high performance is closely related in this case with high resource requirement and use. The recorded best performing method is also by far the one requiring the most resource, including the vector size, among other things as already mentioned.

It, therefore, stands out that though FastText performs better than other embeddings, it does so only marginally and also require more resources (Figure 11.6).

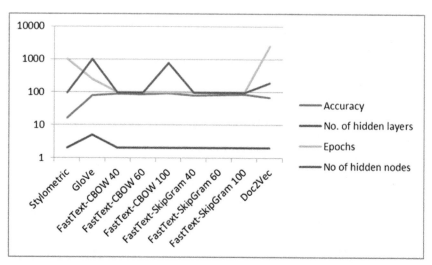

FIGURE 11.6 Comparative results across approaches.

KEYWORDS

- **artificial neural networks**
- **authorship identification**
- **common bag of words**
- **global vectors**
- **neural embeddings**
- **word embeddings**

REFERENCES

1. Seroussi, Y., Zukerman, I., & Bohnert, F., (2011). Authorship attribution with latent Dirichlet al.ocation. *CoNLL '11: Proceedings of the Fifteenth Conference on Computational Natural Language Learning*, 181–189.
2. Soares, D. O. L., Pavelec, D., & Justino, E., (2007). Author identification using stylometric features. *Inteligencia Artificial: Revista Iberoamericana de Inteligencia Artificial* (No. 36, pp. 59–66). ISSN: 1137-3601 (Ejemplar dedicado a: From Natural Language Processing to Information and Human Language Technology). 11. 10.4114/ia.v11i36.892.

3. Khomytska, I., & Teslyuk, V., (2020). Statistical models for authorship attribution. In: Shakhovska, N., & Medykovskyy, M., (eds.), *Advances in Intelligent Systems and Computing IV. CCSIT 2019* (Vol. 1080). Springer, Cham.
4. Rexha, A., Kröll, M., Ziak, H., & Kern, R., (2018). Authorship identification of documents with high content similarity. *Scientometrics, 115*, 223–237. https://doi.org/10.1007/s11192-018-2661-6.
5. Chen, Q., Tianchang, H., & Rao, Z., (2017). *Deep Learning-based Authorship Identification*. Accessed at: https://web.stanford.edu/class/archive/cs/cs224n/cs224n.1174/reports/2760185.pdf (accessed on 20 February 2021).
6. Akimushkin, C., Amancio, D. R., & Oliveira, O. N. Jr., (2017). Text authorship identified using the dynamics of word co-occurrence networks. *PLoS One, 12*(1), e0170527. doi: 10.1371/journal. pone.0170527.
7. Nirkhi, S., Dharaskar, R., & Thakare, V. M., (2015). An experimental study on authorship identification for cyber forensics. *International Journal of Computer Science and Network, 4*(5), 756–760.
8. Bozkurt, I. N., Baghoglu, O., & Uyar, E., (2007). Authorship attribution. In: *22nd International Symposium on Computer and Information Sciences* (pp. 1–5). Ankara. doi: 10.1109/ISCIS.2007.4456854.
9. Mohsen, A. M., El-Makky, N. M., & Nagia, G., (2016). Author identification using deep learning. In: *15th IEEE International Conference on Machine Learning and Applications (ICMLA)* (pp. 898–903).
10. Zhou, L., & Huafei, W., (2016). *News Authorship Identification with Deep Learning*. Accessed at: https://www.semanticscholar.org/paper/News-Authorship-Identification-with-Deep- Learning-Zhou-Wang/d2af63c41d164a1f33873150758d427d8d89421b (accessed on 20 February 2021).
11. Le, Q., & Mikolov, T., (2014). Distributed representations of sentences and documents. In: *ICML'14 Proc. 31st International Conference on International Conference on Machine Learning,* (Vol. 32, pp. 1188–1196).
12. Tomas, M., Ilya, S., Kai, C., Greg, C., & Jeffrey, D., (2013). Google Inc. Distributed representations of words and phrases and their compositionality. *NIPS'13 Proceedings of the 26th International Conference on Neural Information Processing Systems* (Vol. 2, pp. 3111–3119).
13. Piotr, B., Edouard, G., Armand, J., & Tomas, M., (2016). *Enriching Word Vectors with Sub Word Information.* arXiv:1607.04606 [cs.CL].
14. Pennington, J., Socher, R., & Manning, C. D., (2014). *GloVe: Global Vectors for Word Representation.* https://nlp.stanford.edu/pubs/glove.pdf (accessed on 20 February 2021).
15. Lewis, D. D., et al., (2004). Rcv1: A new benchmark collection for text categorization research [J]. *Journal of Machine Learning Research, 5*, 361–397.
16. UCI., (2011). *Machine Learning Repository.* https://archive.ics.uci.edu/ml/datasets/Reuter5050 (accessed on 20 February 2021).
17. Dugar, T., & Gowtham, S., & Chakraborty, U., (2019). *Hyperparameter Tuning for Enhanced Authorship Identification Using Deep Neural Networks,* 206–211. 10.1109/ICOEI.2019.8862631.

CHAPTER 12

FUSION-BASED LEARNING APPROACH FOR PREDICTING DISEASES IN AN EARLIER STAGE

K. KRISHNA PRASAD,[1] P. S. AITHAL,[2] A. JAYANTHILADEVI,[1] and MANIVEL KANDASAMY[3]

[1]*Computer Science and Information Science, Srinivas University, Mangalore, Karnataka, India, E-mail: krishnaprasadkcci@srinivasuniversity.edu.in (K. K. Prasad)*

[2]*Srinivas University, Mangalore, Karnataka, India*

[3]*Optum Tech-United Health Group, Bangalore, Karnataka, India*

ABSTRACT

With the recent enhancement encountered in healthcare systems, the total amount of healthcare data raised drastically in diverse factors. These sorts of data are generated from diverse sources such as mobile, digital fields, and wearable devices. Big data may provide suitable opportunities for data analysis and improvement in healthcare-based services through rising technologies. The ultimate objective of this investigation is to construct a structural mathematical model to improve disease prediction with fused nodes using learning approaches. This node is based on an information model for developing certain prediction systems. Learning may co-operate various difficult approaches like a neural network, a Bayesian model for extracting data, and logical inference. This learning model may combine information paradigm, which can be cast off to offer a reliable and comprehensive prediction model for healthcare data. With this anticipated model, an experimental analysis with mathematical illustration is provided here. The analysis is done with a MATLAB environment for projecting the functionality of the anticipated model.

12.1 INTRODUCTION

The ability of digital healthcare services is to change the complete healthcare process to be more appropriate and effective patterning. The generalized form of digital healthcare models includes mobile devices, health records, and wearable health devices.

Initially, these health records are sourced from check-ups by patients and diagnosing patient's data. Digitalization may offer health-based record sharing over diverse organizations. With these records, physicians may show superior concern towards the medical history of patients. Moreover, when data gather over time, these records may gather in larger volumes. This leads to complexity in storing, processing, and retrieval. Certain estimation based on these records provides higher information. With association with mobiles, apps that are accessible for numerous functionalities like assessment, decision support system, practical management, treatment, and care. Wearable devices may provide the fastest growth in changing conventional health care activities and constant health management. Various investigations may estimate wearable devices may attain global functionality. Sensors embedded in those devices may facilitate attainment of healthcare data. For instance, blood pressure and heart rate can be identified using smartphones. With rapid growth in these data, investigators may determine improved value to merge learning and fusion for examining huge amount of data. Learning facilitates machine learning (ML) models at various levels to be cast-off diverse non-linear functional layers. Fusing information may constantly facilitate attainment of appropriate data to acquire awareness that may offer decision making functionality.

With faster growth towards data, investigators may determine the rising value of learning and fusing information for examining enormous amount of data. This learning model may utilize continuous procedures to acquire appropriate information to attain essential awareness for assisting decision making support system. With advent healthcare approaches like age associated crisis and chronic diseases, investigators are constantly acquiring solution for predicting diseases and diagnosis in a more appropriate manner.

With fusion approaches using healthcare data to offer viable outcomes has turned to be prime industrial topics. This investigation attempts to offer an improved structural model for constructing risk prediction model with integration of learning and fusion-based approaches.

Various investigators have started to analyze diverse perspectives of heath data from diverse available repositories and to make appropriate decision for predicting disease. Moreover, constant development shows an ability to maintain infancy. Here, overall functionalities and approaches are provided for modeling and examining health care data that are more appropriately developed. Traditional data mining techniques may encounter drawbacks with efficiency and accuracy owing to its constraints towards data processing and quality. Numerous traditional models may feel complexity in examining data context. In certain environments, this idea is considered to be extremely complex. Specifically, when it undergoes functional dependencies with complex nature, it shows an inability to express data in a simpler manner. The industrial model may offer advanced approaches that facilitate information extraction from unstructured data in larger volumes.

The remainder of the work identifies the research ability of learning and health data. Section 12.2 offers proposed methodologies; Section 12.3 explains numerical results and discussions, and Section 12.4 depicts the conclusion of the anticipated model and future work directions.

12.2 METHODOLOGIES

Healthcare industries have gained potential advantages gained from data analytics. It has extensive investigation of architectural model and execution of data in health care industries.

12.2.1 ARCHITECTURAL DESIGN

The essential need of data analytics is to handle variety, volume, and data velocity that commences from source for sharing context. Data will be generated and stored in various sources like relational databases, documents, XML, and so on. Changing these datasets towards sharable and understandable form needs services that have to be collected, processed, and prepared from those datasets. Architectural modeling of big data in healthcare is like a conventional data analytics model that performs references towards conventional data analytics. There are diverse attempts that have been made towards healthcare domains. This system comprises of five layers:

1. data;
2. aggregation;
3. analytics;
4. exploration; and
5. data governance.

12.2.2 FUSION MODEL

By merging this differentiation, fusion may carry out mining to dimes uncertainty and acquire superior functionality of information. By adopting these statistics, this work may offer numerous approaches for fusing information. For instance, a Bayesian classifier is used. This fusion approach is cast-off to gain inference regarding event identity in observation space. This may use probability distribution for fitting the model towards unobserved and observed data. It process may be applied to merge information sources. While processing inference, the Bayesian fusion procedure for all data sources may offer a hypothetical observation and source-based process. Hypothesis like $H^n (k=1,2,...n)$ is cast-off to compute probability of every entity with functionality $P(H^k|O_i)$, where 'i' specifies data types and O_j is data source entity. Data source probability may be merged with Bayesian inference. Output may be merged with probability of $P(H^1, H^2,...,H^n|O_i)$. This logic is cast to optimize merged probability when there are constraints on last outputs. This procedure is termed as fusion identity. Ground truth of inference is based on rules with probability grounded on reasoning. This may be computed with hypothetical probability on event which is provided as in Eqn. (1):

$$P(H|E) = \frac{P(H,E)}{P(E)} \qquad (1)$$

where; $P(H,E)$ is intersection probability of hypothesis and event 'E' This classifier inference is provided as in Eqn. (2):

$$C_i = \arg\max_{c_i} P(C_i \mid x) \qquad (2)$$

where; $C_i\,(i=1,2,\ldots,m)$ specifies 'm' class set, $P(c_i \mid x)$ is posterior probability and 'x' is unidentified feature vectors. These attributes are provided based on its class for all its maxima probability. This rule may facilitate appropriate computation of prior probability for diverse unknown event when probability evidence regarding event are determined. Assumptions are known based on unknown event and proofs are independent. Moreover, there are numerous unknown events and diverse evidence. Henceforth, this work is initiated to specify joint probability between provided variables by DAG, where nodes may specify directed edges and random variables which specifies dependencies among variables. This structural model is termed as DAG. Joint probability is depicted as in Eqns. (3)–(5):

$$P(U) = P(X_1,\ldots,X_n) \qquad (3)$$

$$P(U) = \prod_{X_i \in U}^{1} P(X_i \mid X_1,\ldots,X_n) \qquad (4)$$

$$P(U) = \prod_{i=1}^{n} P(X_i \mid P_a(X_i)) \qquad (5)$$

where; $U = \{X_1,\ldots,X_n\}$ is random variables set termed as universe and $P_a(X_i)$ is parental variables of DAG. Nodes in graph to factor $P(X_i \mid P_a(X_i))$. This network is a fusion method that is more appropriate for uncertainty measurement for property extraction with graphical structural and calculus probability. Association among these models are $\{X_1,X_2,\ldots,X_M\} = \{X_{m+1},X_{m+2},\ldots,X_n\}$ where nodes specify edges and variables in direct specification. Probability distribution of network is depicted as in Eqns. (6)–(9):

$$P(X_1,X_2,\ldots,X_m X_{m+1},\ldots,X_n) = \prod_{i=1}^{n} P(X_i \mid p(X_i)) \qquad (6)$$

$$P(X) = \prod_{i=1}^{n} P(X_i) \prod_{j=m+1}^{n} P(X_j \mid X_1,\ldots,X_m) \qquad (7)$$

$$P(X) = \prod_{i=1}^{n} S(\{X_i\}) \prod_{j=m+1}^{n} \frac{P(\{X_j, X_1, \ldots, X_m\})}{P(\{X_1, X_2, \ldots, X_m\})} \quad (8)$$

$$P(X) = \prod_{i=1}^{n} S(\{X_i\}) \prod_{j=m+1}^{n} \frac{S(\{X_j, X_1, X_2, \ldots, X_m\})}{S(\{X_1, X_2, \ldots, X_m\})} \quad (9)$$

Here, 'S' is support confidence of item set. Fusing can be based on feature selection that is extremely complex in various approaches. For instances, this recognition task over noisy circumstances is complex owing to interference of inappropriate features. These approaches can be cast-off to haul out essential features (Figure 12.1).

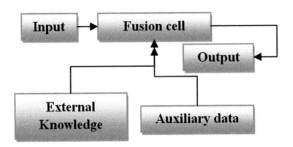

FIGURE 12.1 Fusion model.

An associate neural network is cast-off for categorizing data types. This network is trained to generate identity approximation-based mapping among inputs/outputs with the back-propagation model. Network with i/o nodes and hidden layers with linear activations. In training model, the projection matrix is computed with input dataset for normalization. Output from network model is provided to neurons. In this non-linear process, tangent process is known as the activation function. This function is depicted as in Eqn. (10):

$$z = A(u) = 1 - \frac{2}{1 + \exp(2u)} \in (-1, 1) \quad (10)$$

where; 'u' is specified as input dataset and 'Z' is dataset output. This may be compared with appropriate outputs of all these data types, weights may be adjusted for reducing errors. After training, weights may be preserved on fusion modeling. As data types are determined, format specification is provided as external specification that is utilized to validate identification outcomes, meta-data extraction of object files are normalized based on object content in common specification.

12.3 NUMERICAL RESULTS

For executing Bayesian rules, feature extraction-based selection is cast-off for accessing libraries. Simulation has been done in MATLAB environment. For modeling these networks, regression model for time series analysis in state estimation components may use libraries. Benefits of using file support and volume for higher efficiency for appropriate data access. It is utilized for handling raw data and data alignment. It may offer fault tolerance in handling huge data in sparse data. Communication among data and system is through services executed in MATLAB. To validate selection performance of CNN selection execution, this may execute CNN with learning model in SVM, k-NN for comparison. Health records may be roughly 20,000 were samples for computation. Original records are based of sample members. There may be duplicate records that are generated from source systems. For instance, users may upload similar data for numerous times. Subsequently, invalid records may prevail owing to malfunction systems or inappropriate software operations. Data quality is needed for fulfilling the performance and accuracy of analytics. This is attained by executing filters in data alignment that may screens duplication and outlier for processing. Accuracy and kappa measures for certain execution are provided in Figure 12.2.

Tables 12.1 and 12.2 depict the overall disease prediction model based on accuracy. Various diseases are analyzed here—diabetes, obesity, stroke, heart disease, cancer, and so on.

TABLE 12.1 Accuracy Computation

Methods	SVM	k-NN	CNN
Duration (Min)	1248	890	1600
Throughput (r/sec)	4148	5700	3200
Throughput (KB/sec)	3380	4600	2620

TABLE 12.2 Overall Disease Prediction

Risks	SVM	k-NN	CNN
Diabetes	80	66	97
Obesity	54	85	72
Heart disease	74	85	73
Stroke	66	68	48
Blood pressure	69	53	76
Cancer	61	60	84
Overall accuracy	68	66	75

Figures 12.2–12.4 depict performance measures of the anticipated approach. Here, SVM, k-NN, and CNN models are used for comparison.

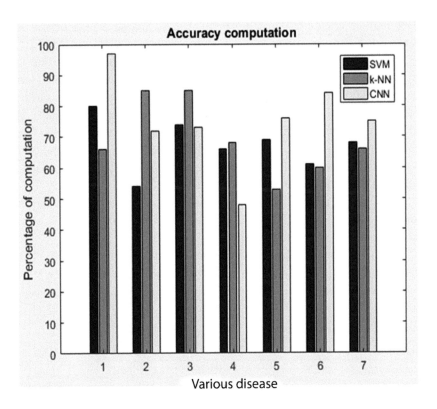

FIGURE 12.2 Accuracy computation.

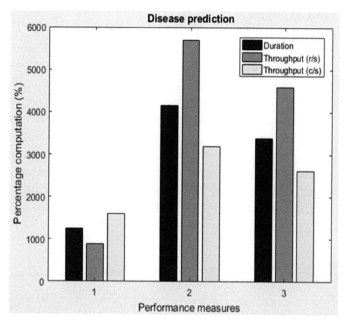

FIGURE 12.3 Disease prediction.

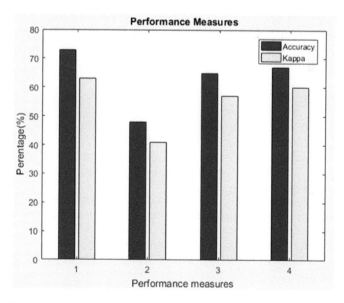

FIGURE 12.4 Performance measures.

12.4 CONCLUSION

This investigation utilizes potential competency for using fusion concept and learning model to improve healthcare analytics, therefore this process may be done with effectual computation. In ML process, complex functionality has been done for enhancing analytical performance. With this learning process, it may facilitate risks to identify prediction in a more appropriate manner. Here, network model has the ability to improve analysis for training accuracy. This specification has enormous training sets. This is because of analytical components and workflow. The architectural model is based on anticipated framework to organize components of the analytical system to acquire fusion. In the future, the MapReduce concept will be used for enhancing the approach.

KEYWORDS

- Bayesian approach
- Big data
- learning model
- neural network
- structural modelling

REFERENCES

1. Deng, L., & Yu, D., (2013). Deep learning: Methods and applications. *Foundations and Trends in Signal Processing, 7*(3/4), 197–387.
2. Wang, Y., Kung, L., & Byrd, T. A., (2016). Big data analytics: Understanding its capabilities and potential benefits for healthcare organizations. *Technological Forecasting and Social Change*, 1–11.
3. Wang, Y., Kung, L., Ting, C., & Byrd, T. A., (2015). Beyond a technical perspective: Understanding big data capabilities in health care. In: *Proceedings of the 48th Hawaii International Conference on System Sciences* (pp. 3044–3053).
4. Mitchell, H. B., (2012). *Data Fusion: Concepts and Ideas*, 1–344.
5. Bosse, E., & Solaiman, B., (2016). *Information Fusion and Analytics for Big Data and IoT*. Artech House.

6. Wang, C., & Zeng, D. D., (2016). Mining opinion summarizations using convolutional neural networks in Chinese microblogging systems. *Knowledge-Based Systems, 107*, 289–300.
7. Abdel-Hamid, O., Mohamed, A. R., Jiang, H., Deng, L., Penn, G., & Yu, D., (2014). Convolutional neural networks for speech recognition. *IEEE Transactions on Audio, Speech and Language Processing, 22*(10), 1533–1545.
8. Allard, D., Comunian, A., & Renard, P., (2012). Probability aggregation methods in geoscience. *Mathematical Geosciences, 44*(5), 545–581.
9. Tian, D., Gledson, A., Antoniades, A., Aristodimou, A., & Dimitrios, N., (2013). A Bayesian association rule mining algorithm. In: *Proceedings of the IEEE International Conference on Systems* (pp. 3258–3264).
10. Vermeulen-Smit, E., Ten, H. M., Van, L. M., & De Graaf, R., (2015). Clustering of health risk behaviors and the relationship with mental disorders. *Journal of Affective Disorders, 171*, 111–119.

CHAPTER 13

A FUZZY-BASED FRAMEWORK FOR AN AGRICULTURE RECOMMENDER SYSTEM USING MEMBERSHIP FUNCTION

R. NARMADHA,[1] T. P. LATCHOUMI,[2] A. JAYANTHILADEVI,[3] T. L. YOOKESH,[4] and S. PRINCE MARY[5]

[1]Department of ECE, Sathyabama Institute of Science and Technology, Chennai, Tamil Nadu, India

[2]Department of CSE, VFSTR (Deemed to be University), Andhra Pradesh, India

[3]Computer Science and Information Science, Srinivas University, Mangalore, Karnataka, India, E-mail: drjayanthila@gmail.com

[4]Department of Mathematics, VFSTR (Deemed to be University), Andhra Pradesh, India

[5]Department of CSE, Sathyabama Institute of Science and Technology, Chennai, Tamil Nadu India

ABSTRACT

The primary key factor that has been focused in this work is 'agriculture.' Indian farmers have knowledge and approach for cultivating and growing crops based on land nature that has evolved around thousands of years of observation and experience, and that is the main cause to explain why farmers produce food without such infrastructure for over billions of people in every year. This fuzzy-based model may predict items that are consumed by every customer; therefore, farmers may produce items based on their essential choices.

13.1 INTRODUCTION

The primary key factor that has been focused in this work is 'agriculture.' Agriculture provides 56–60% of countries employment; however, outcomes in merely 18% of GDP [1]. India is one amongst the world that considers 4–5 crops a year. Indian farmers have knowledge and approach for cultivating and growing crops based on land nature that has evolved around thousands of years of observation and experience, and that is the main cause to explain why farmers produce food without such infrastructure for over billions of people in every year [2]. However, in present time farmers encounter enormous confronts such as competition, whether from the huge corporation and rise in price creating enormous economic pressures that cause considerable distress [3]. Climate variation may lead to distorted farmers' economy by 25–30% in shorted month span based on the Indian government's latest annual economic survey performed [4]. Report shows that extreme shocks degree may reduce farmers' monthly income by 4.5%, and heavy rainfall may reduce their income by 14%.

In recent times, virtual resources can be accessed from anywhere and everywhere with more comfort is known as cloud [5]. This technology assists in accessing diverse virtual resources to utilize diverse services and application through World Wide Web and certain standard protocols for communication termed as cloud computing [6]. CC offers dynamic resources and makes the users to pay per uses billing policies and metering. The foremost cloud benefits are its flexibility of resources utilization anywhere and anytime [7]. CC provides computational power like CPU, RAM, network speed, OS, storage as services over network indeed of providing computational resources at users' place [8]. Hardware usage and software may diminish by utilization of this CC technology at users' side. With the assistance of this simple web-based browser, users may run CC systems. However, the rest will be handled by the cloud-based network.

For overall countries development, which includes agricultural sector like CC as an essential role to be play? Agriculture sector is provided by effectual execution of CC [9]. In the modern time period of Information technology, CC technology may be extremely essential for the centralization of every agriculture-based database. CC utilization in agriculture may offer data readiness anywhere and anytime which may improve the GDP of the nation [10]. It will also provide food security level and also offers communication globally and locally.

Recommender system or recommendation system is an information sub-class that recommends content to users and may interpreted based on aggregated data libraries. It is considered to be a system that may filters information that identifies 'preference order' or 'rating' of individual items that may be provided by users [11]. The ultimate objective of recommendation system is to offer essential recommendation to users of group for all products or items which will be more inclined to be forwarded. RR may be utilized in various fields like movies, music, books, research articles, clothes, news, and so on [11]. It may reduce information overloading complexity by filtering essential information out of dynamical amount that are processed information based on its interest and preferences of users regarding item set. With respect to users profile, this system may predicts whether a specific user may prefer any sort of items or not [12]. It is also validated to enhance quality and decision-making process (Figure 13.1).

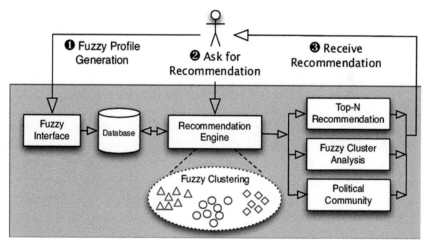

FIGURE 13.1 Generic fuzzy-based recommender system.

Certain kind of recommendation systems are more appropriate to users either in morning or in the evening, and he/she may be alike to carry of performance when it may be either hot or cold outside however only on certain conditions [13]. Recommendation system may be provided with huge attention and utilizes these kinds of details or information in offering recommendation is termed as context-aware recommendation systems [14]. This work utilizes fuzzy-based approach for effectual recommendations. This work utilizes numerous criteria like soil, crop type, water required, soil type, sunlight intensity needed, nutrients, soil fertilizers and pH level for effectual and efficient recommendations. This sort of recommendation may be provided based on factors like crop or soil which is not so appropriate for effectual recommendation.

13.2 RELATED WORKS

Enormous studies have been performed on cultivation in the Nile delta region; however, few investigators have handled the cultivation in desert regions, specifically oasis. However, extremely limited investigations studies have been anticipated for computational intelligence-based recommender system for analyzing effectual planting dates specifically in case of newly cultivated lands with respect to essential temperature for stages of crop growth. For instance, the author in Ref. [15] anticipated a recommender system based on rough metrological theory for analyzing finest cultivation dates for wheat over the Egyptian Sinai peninsula based on essential mean temperature for germination phase.

There is a huge investigation that uses the idea of an expertise system to assist farmers for enhancing their cultivation practices [16]. Numerous expert systems in the past may concentrate on rising productivity by choosing plants that are more appropriate to land, like climatic conditions and soli suitability in southern Greece [17] using computation based on choosing crops and land suitability that has to be planted which maybe olives, wheat, grapes, wheat, and tomatoes. Numerous studies have been concentrated based on farmers recommendation systems [18]; for instance, enhancing soybean yields by recommending optimal decision making for farmers and utilization of inorganic fertilizers for sugarcane. Numerous systems were modeled for predicting plant disease like soybean disease and oil crop-based diseases.

Numerous approaches have been utilized like recommendation-based services; for instance, rule-based, ontology-based, and fuzzy decision-based approaches. There are diverse factors that resemble these works in

Ref. [19]. Moreover, this investigation is more dedicated in choosing rice varieties. To be more specific in suitable factors, which may not be exclusively measured with factors associated with climate and land, that may consider the probability of spreading diseases and pests, planting periods, and risk from natural disasters.

This work offers a scheme for choosing agriculture-based knowledge by formulating rule based on Fuzzy and membership function (MF) along with implementation of more specific recommendation system that recommends rice varieties that are more appropriate for planting periods and land and planning crop by offering personalized crop-based calendar which assists in certain methods as crop manager in Ref. [20].

At present, with utilization of modern information technology to construct decision-based support system for precision fertilizer for modifying conventional fertilization custom in hotspot research field in precision-based fertilization [21]. Numerous investigators have developed and investigated GIS-based precision fertilizer management system and decision-making approach which considers farmland as objects; however, eliminating farmer's status as essential farming in China-based villages [22]. Development system has restricted amount of regional application that may causes difficult management, high pollution, and higher cost and complex popularization. Therefore, it is more appropriate to realize a simple, practical, and expansible system for reducing system complexity, diminishing crop production-based cost, strengthening pervasiveness system to improve agricultural production efficiency, enhancing agricultural ecological circumstances and improve comprehensive agriculture production ability.

With the improvement in modern information technology, GIS server is measured to be of complete, lower cost and effectual process [23]. This work considers crop fertilization-based decision-making system at administrative village scaling like research papers that gathers soil information and map-based village level using views; then model a crop-based recommendation system for fertilization with available servers and executed in farmlands of village level with nutrient management and recommending online fertilizers [24]. It may offer online soil-based information query and fertilization-based decision-making system and offer technical assistance for scientific fertilization.

13.3 METHODOLOGY

Recommender system may function specifically with two essential approaches based on the content-based and collaboration-based models. This effectual and simple procedure for generating rule discovery is done with Fuzzy and membership modeling that is considered to be a design model. Active customers have established certain agreements with customers in the past, then this recommendation system will provide an essential relevance and interest towards active users.

The necessity of filtering the complete product range for accessible alternatives may produce diverse recommendation more necessarily. It is considered to be extremely easier choice of items to be superior; however, too much choice is not so suitable. RS may produce recommendation that may assist customers' data, diverse knowledge, historical data, and available item regarding transactions with references to that database. Customers can surf through these RS and reject or accept various items; therefore, feedback is more essential. Every transaction is stored in the database, and this newer recommendation system is produced during successive item-based ordering.

Development in RS may utilize customer-based and item-based models. In item-based approaches, there are numerous products that are arranged using customers that remains in lesser or more similar. Certain items may assist in constructing corresponding neighborhoods based on appreciations attained from diverse customers. Later, these systems may produce recommendation based on diverse items, customers may prefer those items. In customer-based model users play a dominant role. Customers may order similar items that are grouped together. System model is considered to be a hybrid that merges these content and collaborative fuzzy approaches. Figure 13.2 depicts the architecture model of agriculture-based recommendation system.

These kinds of systems may be modeled as web-based graphical user interface and based on mobile application. Complete requests that are generated were works based on this webserver. Every input will be stored in hybridization system and database which may provide appropriate recommendation in prediction form of the purchase of diverse items that has to be modeled.

A Fuzzy-Based Framework for An Agriculture Recommender System

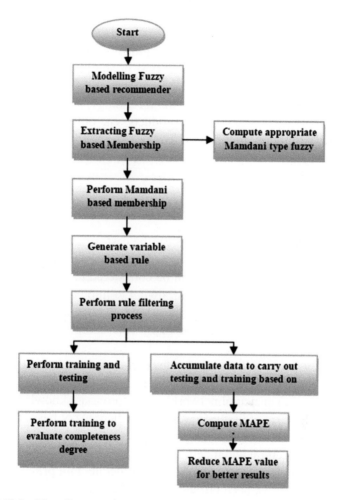

FIGURE 13.2 Flow diagram of proposed model.

By considering all these items that are ordered by every customer, cumulative order of these items by every customers assist in predicting usage of certain item at certain cause of year. Therefore, this system may assist in making prediction of consumption of diverse customer products. Therefore, farmers may also make vegetables production and fruits based on RS.

This model cast of MF and fuzzy model for executing and designing RS. India is considered to be an agriculture-based country that uses conventional way of recommenders in case of agriculture sector. At present, recommendations, and knowledge are provided via interaction among experts and farmers. These kind of recommendations may be diverse for all diverse experts. This model may assist in attaining recommendations and information dependent on their preferences and needs. This may help to diminish the gap among customers, technology, and farmers. Farmers and customers can access these products via online websites. Some android based mobile applications are also used, and some apps can be downloaded for attaining recommendations in the essential time period.

13.3.1 FUZZY-BASED RECOMMENDATION SYSTEM (FRS)

In an extensive broader sense, FRS can be cast-off as a basic for various kinds of system knowledge, or to model relationships and interactions between system variables. FRS can be considered to be an essential tool for modeling these complex system; owing to imprecision or complexity where conventional tools are not successful. A huge amount of methods have been anticipated to produce fuzzy rule automatically from those numerical data.

Generally, it may utilize complex fuzzy rules from numerical data. They utilize complex rule-based generation for diverse approaches like genetic algorithms (GAs) or neural networks. An appropriate knowledge may be modeled in the form of linguistic variables, where fuzzy-based MF parameters like fuzzy rules, number of rules, and so on. Generic code structure may provide structural and independent feature-based performance that is more appropriate for candidates that in co-operate with prior knowledge. Various benefits have been provided that utilizes MF for modeling Fuzzy based recommender system. Knowledge base on FRS comprises of two components: rule base and data base. Rule base comprises of set of fuzzy rules while database comprises of scaling factors and holds MFs of fuzzy set that specifies linguistic terms.

Membership functional shape is based on two diverse variants based on fuzzy model whether descriptive and approximate. Fuzzy model is considered to be a more qualitative system expression. KB is a fuzzy set which is considered to be more semantic to linguistic labels which defines rules, including KB. This system may use global semantics. In approximate fuzzy modeling, KB is determined to be a fuzzy rules to provide own meaning, i.e., linguistic variables comprises of rules that are not consider its value with labels from global term set. At last, linguistic variables are considered to be fuzzy variables (FVs). Approximate fuzzy mode is more precise with descriptive model. This work considers Mamdani type fuzzy-based rule model for agriculture-based recommendation system. In Mamdani type, general rules are provided as:

If x_1 is a_i and x_2 is a_j, then 'y' is b_j, where x_1, x_2, and y are linguistic variables and a_i, b_i are fuzzy sets. Here, initially, DB is used for uniform fuzzy partitions that use triangular MFs with 0.5 height consideration. Number of linguistic terms specified by experts where there is necessity to certain primary fuzzy set that belongs to everyone in fuzzy partitions. FVs x_i is primary set $t(x_i) = \{l_1(x_1), l_2(x_2)..., l_n(x_i)\}$ with related ordered set $t'(x_i) = \{1, 2, ..., n\}$. Subsequently, second art is triangular MFs with rules, $l_i(x_i)$ by association of tuples $\{al_i, bl_i, cl_i\}$.

1. Use prevailing primary linguistic fuzzy-based partitions and produce individuals by considering rule cover for randomly attuned samples. Primary fuzzy-based sets and meaning are provided here.
2. Generate individual in a similar way, compute values randomly and produce gene variation with respective intervals.
3. Produce model with sample rule generation.

$$c_1^{t+1} = ac_w^t + (1-a)c_v^t \qquad (1)$$

$$c_2^{t+1} = ac_v^t + (1-a)c_w^t \qquad (2)$$

$$c_3^{t+1} \text{ with } c_{3k}^{t+1} = \min\{c_k, c_k'\} \qquad (3)$$

$$c_4^{t+1} \text{ with } c_{4k}^{t+1} = \max\{c_k, c_k'\} \qquad (4)$$

From the above equation, it is well known that operators may use parameter that is either variable or constant whose value may base on available variables. Similarly, a combination of two diverse MFs may depend on diverse fuzzy set that makes two newer fuzzy set to belong in intervals of performance determination by initial fuzzy partition.

$$c_t = \begin{pmatrix} c_1,..,c_p,c_{p+1},...,c_{n+1},ac_1,bc_1,cc_1,. \\ .,a_{c_p},b_{c_p},c_{c_p},...,a_{c_{n+1}},b_{c_{n+1}},c_{c_{n+1}} \end{pmatrix} \quad (5)$$

Here, three correlation parameters are considered (x_1, x_2, x_3) that describes triangular MF and $x_1 \leq x_2 \leq x_3$ has to be confirmed with the certain fuzzy set. With $c_i = (x_0, x_1, x_2)$ are MF that are adopted, related performance interval is provided as $\left[c_i^1, c_i^2\right] = [x_o - \frac{x_1 - x_0}{2}, x_2 - \frac{x_2 - x_1}{2}]$. This adoption is based on fuzzy set generation $C_i' = (x_0', x_1', x_2')$ by modal point, respectively.

Similarly, covering approach is modeled based on iterative approaches that facilitate to acquire fuzzy rule set that covers sample values. In every iteration, it works on production approach, that acquires finest fuzzy rue based on current training set state, it may consider relative value for provoking rules over it, and eliminates samples with coverage value that are higher than ε. Covering value is provided with equation given below:

$$CV(e_l) = \sum_{l=1}^{p} R_i(e_l) \quad (6)$$

13.3.2 RULE FILTERING

The primary variable is generated from initial KB. This may be encoded directly to filter rules that are specified as c_j remaining variables are produced by relating interval of performance $\left[c_h^l, c_h^r\right]$ to all variables in $c_{1,h} = 1,2,..,3NM$. Every performance interval will be adjusted based on corresponding variables like $c_h \in \left[c_h^l, c_h^r\right]$. If $t_{mod} 3 = 1$, then c_t is value left in support with triangular fuzzy membership. This triangular fuzzy model is provided by three factors (c_t, c_{t+1}, c_{t+2}) and performance interval is provided as in equation below:

$$c_t \in \left[c_t', c_t^r\right] = [c_t - \frac{c_{t+1}}{2}, c_t + \frac{c_{t+1} - c_t}{2}] \quad (7)$$

$$c_{t+1} \in \left[c_{t+1}^l, c_{t+1}^r\right] = \left[c_{t+1} - \frac{C_{t+1} - c_t}{2}, c_{t+1} + \frac{C_{t+2} - c_{t+1}}{2}\right] \quad (8)$$

$$C_{t+2} \in \left[C_{t+2}^l, C_{t+2}^r \right] = \left[C_{t+2} - \frac{C_{t+2} - C_{t+1}}{2}, C_t + \frac{C_{t+3} - C_{t+2}}{2} \right] \quad (9)$$

13.4 NUMERICAL RESULTS AND DISCUSSIONS

Based on the above computation, simulation has been done in MATLAB environment. This recommender model based on fuzzy MF provided better trade-off in contrary to other approaches. This helps in generating rule and to make decision which may be applied in agricultural field to assist farmers for making proper cultivation.

Here, extraction of Mamdani type fuzzy-based rule system, rule filtering may construct every cluster for training data. At last, in the testing stage, data are provided for predicting crop yield.

Here, mean absolute percentage error is computed with Eqn. (10) given below:

$$MAPE = \frac{\sum_{i=1}^{r} \left(\frac{abs(p_i - y_i)}{y_i} \right)}{r} X 100\% \quad (10)$$

Tables 13.1 and 13.2 explain about the data gathered and stored in the cloud for making an appropriate prediction with a Fuzzy recommender system for crop cultivation. These data have been collected from the duration of June 2019 to February 2020 for interpretation. Based on these collected data, Table 13.3 depicts the mean error rate computation. The anticipated Fuzzy model shows lesser error in contrary to other approaches like ANN, GA, PSO, hybrid fuzzy, and so on.

TABLE 13.1 Agriculture-Based Data Extraction for Fuzzy Training

Training Data		
From	To	Total Amount
June 2019	December 2019	500/–

TABLE 13.2 Agriculture-Based Data Extraction for Fuzzy Testing

Testing Data		
From	To	Total Amount
December 2019	February 2020	100/–

TABLE 13.3 MAPE Computation

Techniques	MAPE Values
HMM	1.220
ANN and GA	0.850
Hybrid Fuzzy	0.780
ARIMA	0.902
ANN	0.902
PSO	0.57
Fuzzy with Mamdani	0.39

Figures 13.3 and 13.4 depict the Mamdani-based MF for an effectual recommender system. Figure 13.5 explains about outcome that has been attained after performing membership operation. With filtered rules, these outcomes are computed. Figures 13.6 and 13.7 provide a graphical model of training and testing agricultural data that has been used by the recommender system. Figure 13.8 demonstrates the graphical model of MAPE computation. From all these outcomes, it is identified that Fuzzy based rule generation model provides the finest functionality in contrary to other models. Thus, it helps in assisting farmers to make proper cultivation during suitable cultivation period.

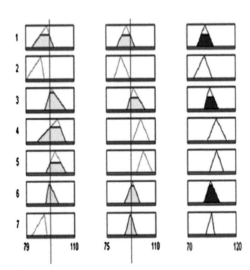

FIGURE 13.3 Mamdani membership function for training data.

A Fuzzy-Based Framework for An Agriculture Recommender System

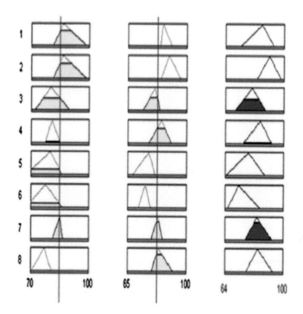

FIGURE 13.4 Mamdani membership function for testing data.

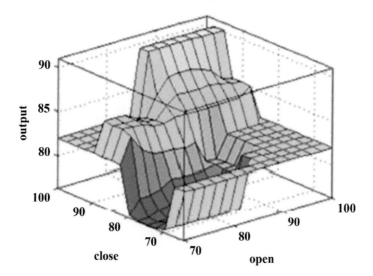

FIGURE 13.5 Fuzzy-based output generation.

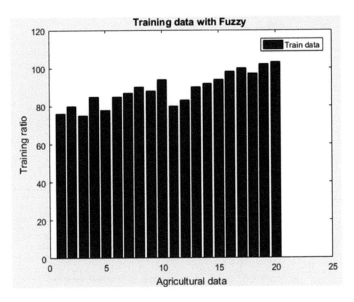

FIGURE 13.6 Training model with fuzzy membership function.

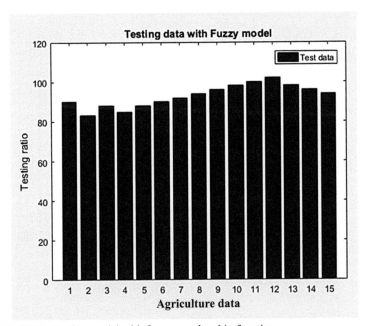

FIGURE 13.7 Testing model with fuzzy membership function.

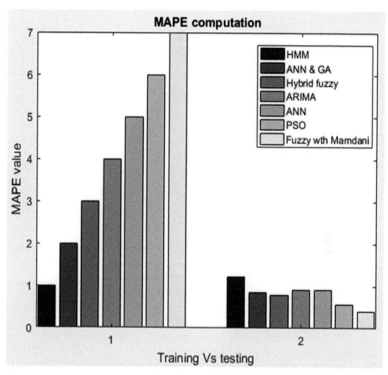

FIGURE 13.8 Graphical representation of MAPE computation.

13.5 CONCLUSION

A recommender system is modeled specifically for recognizing and predicting consumption of diverse agriculture-based items that are modeled and developed with fuzzy-based MFs. This system model was competent of providing prediction and recommendation based on customers' needs by buying characteristics and corresponding peer item-based recommendation. This fuzzy-based model may predict items that are consumed by every customer; therefore, farmers may produce items based on their essential choices. Therefore, cumulative prediction may assist the farmers to plan and produce cultivation of agriculture-based items/products for any season, therefore, it may not lead to any wastage of items provided by farmers. Therefore, system development assists customers to recommend more such items than predicted ones.

KEYWORDS

- data processing
- dimensionality reduction
- fuzzy-based recommendation system
- fuzzy model
- information technology
- recommendation system

REFERENCES

1. Ye, M., Tang, Z., Xu, J. B., & Jin, L. F., (2015). Recommender system for e-learning based on semantic relatedness of concepts. *Information, 6,* 443–453. http://www.mdpi.com/2078-2489/ 6/3/443.
2. Bart, P. K., Martijn, C. W., Zeno, G., Hakan, S., & Chris, W., (2012). Explaining the user experience of recommender systems. *User Model User-Adapt. Interact., 22,* 441–504. http://dx.doi.org/10. 1007/s11257-011-9118-4.
3. Burke, R., (2002). Hybrid recommender systems: Survey and experiments. *User Model User-Adapt. Interact., 12,* 331–370. http://dx.doi.org/10.1023/A:1021240730564.
4. Chien-Yeh, H., Li-Chieh, H., Tzuo, M. C., Li-Fu, C., & Jane, C. J. C., (2011). *A Web-Based Decision Support System for Dietary Analysis and Recommendations.* The University of Taiwan at Taipei Medical. doi: =10.1089/tmj.2010.0104.
5. Chandra, P., Amar, S. R., & Gaur, S. M. T., (2014). *Fuzzy Based Agriculture Expert System for Soybean.* The University of India at Lovely Professional. doi: 10.13140/2.1.1765.0567.
6. Sri, H., & Imas, S. S., (2010). A fuzzy based decision support system for evaluating land suitability and selecting crops. The University of Indonesia at Bogor Agricultural. *Journal of Computer Science, 6*(4), 417–424.
7. Savita, K., Raj, K., Harvinder, S. S., & Gupta, G. K., (2010). A web-based intelligent disease-diagnosis system using a new fuzzy-logic based approach for drawing the inferences in crops. *Computers and Electronics in Agriculture, 76*(1), 16. doi: 10.1016/j.compag.2011.01.002.
8. Savita, K., Raj, K., Harvinder, S. S., & Gupta, G. K., (2013). Expert system for disease diagnosis in soybean-ESDDS. In: *Proceedings of the Journal of the Indian Society of Agricultural Statistics, 67*(1), 79–88.
9. Jadhav, S. K., Yelapure, S. J., & Babar, V. M., (2011). Rule based expert system in the use of inorganic fertilizers for sugarcane crop. *International Journal of Computer Applications 36*(4), 34–42. doi: 10.5120/4477-6291.
10. Attaher, S., Medany, M., & El-Gindy, A., (2010). Feasibility of some adaptation measures of on farm irrigation in Egypt under water scarcity conditions. *Options. Mediterraneennes, 95,* 307–312.

11. Nelson, G. C., Rosegrant, M. W., Koo, J., Robertson, R., Sulser, T., Zhu, T., Ringler, C., Msangi, S., Palazzo, A., Batka, M., et al., (2009). *Climate Change: Impact on Agriculture and Costs of Adaptation* (Vol. 21). Intl. Food Policy Res. Inst.
12. Khalil, F. A., Farag, H., El Afandi, G., & Ouda, S. A., (2009). Vulnerability and adaptation of wheat to climate change in middle Egypt. In: *13th Conference on Water Technology* (pp. 12–15). Hurghada, Egypt.
13. Dath, A., & Balakrishnan, M., (2013). Development of an expert system for agricultural commodities. *International Journal of Computer Science and Applications (TIJCSA)*, *2*(07).
14. D'Orgeval, T., Boulanger, J. P., Capalbo, M., Guevara, E., Penalba, O., & Meira, S., (2010). Yield estimation and sowing date optimization based on seasonal climate information in the three Claris sites. *Climatic Change*, *98*(3/4), 565–580.
15. Wang, J., Wang, E., Feng, L., Yin, H., & Yu, W., (2013). Phenological trends of winter wheat in response to varietal and temperature changes in the north China plain. *Field Crops Research*, *144*, 135–144.
16. Eitzinger, J., Orlandini, S., Stefanski, R., & Naylor, R., (2010). Climate change and agriculture: Introductory editorial. *The Journal of Agricultural Science*, *148*(05), 499–500.
17. Abdelsalam, M., & Mahmood, M. A., (2014). Climate recommender system for wheat cultivation in North Egyptian Sinai Peninsula. In *Proceedings of the 5th International Conference on Innovations in Bio-Inspired Computing and Applications (IBICA2014)* (Vol. 303, pp. 121–130). Ostrava, Czech Republic, Advances in Intelligent Systems and Computing, Springer.
18. Noaman, M. M., (2008). Barley development in Egypt. In: Ceccarelli, S., & Grando, S., (eds.), *Proceedings of the 10th International Barley Genetics Symposium* (pp. 3–15). Alexandria, Egypt, ICARDA.
19. Frank, M., & Mironov, I., (2009). Differentially private recommender systems: building privacy into the net. In: *Proceedings of the 15th ACM SIGKDD International Conference on Knowledge Discovery and Data Mining*. ACM.
20. Francesco, R., Rokach, L., & Shapira, B., (2011). Introduction to recommender systems handbook. In: *Recommender Systems Handbook* (pp. 1–35). Springer, US.
21. Anvitha, H., & Shetty, S. K., (2015). Collaborative filtering recommender system. *Int. J. Emerg. Trends Sci. Technol.*, *2*(7).
22. Olakulehin, O. J., & Omidiora, E. O., (2014). A genetic algorithm approach to maximize crop yields and sustain soil fertility. *Net. J. Agric. Sci.*, *2*(3), 94–103.
23. Serrano-Guerrero, J., et al., (2015). Sentiment analysis: A review and comparative analysis of web services. *Inf. Sci.*, *311*, 18–38.
24. Rushdi-Saleh, M., et al., (2009). Experiments with SVM to classify opinions in different domains. *Expert Syst. Appl.*, *38*(12), 14799–14804.

# CHAPTER 14

IMPLYING FUZZY SET FOR COMPUTING AGRICULTURAL VULNERABILITY

A. JAYANTHILADEVI,[1] L. DEVI,[2] R. KANNADASAN,[3] VED P. MISHRA,[4] PIYUSH MISHRA,[5] and A. MOHAMED UVAZE AHAMED[6]

[1]Computer Science and Information Science, Srinivas University, Mangalore, Karnataka, India, E-mail: drjayanthila@gmail.com

[2]PGCS Department, Muthayammal College of Arts and Science, Rasipuram, Tamil Nadu, India

[3]School of Computer Science and Engineering (SCOPE), Vellore Institute of Technology (VIT) University, Vellore – 632014, Tamil Nadu, India

[4]Department of Computer Science and Engineering, Amity University, Dubai, UAE

[5]Department of Computer Engineering and Applications, GLA University, Mathura, Uttar Pradesh, India

[6]Department of Information Technology, Qala University College, Kurdistan Region, Iraq

ABSTRACT

The necessity of inter-disciplinary factors of collective assessments and drought events with scientists, policymakers, stakeholders, and public participation are measured to be crucial to offer useful, effectual, and novel information management and understanding with drought events. Some regions may rank as lower risk exposed city. Some may locate in lower drought resistibility areas and have higher drought resistibility. Moreover, these universal standards of agricultural risk grade may not prevail over and division methods have to be studied in the future.

14.1 INTRODUCTION

Droughts are measured as socio-environmental factors where hydrological, climatic, environmental, cultural forces, and socio-economic work that may act as a concomitantly to have severity [1]. The necessity of inter-disciplinary factors of collective assessments and drought events with scientists, policymakers, stakeholders, and public participation are measured to be crucial to offer useful, effectual, and novel information management and understanding with drought events [2]. Here, pro-active risk management tries to concert efforts to plan against drought events which possess no doubt based on vulnerability computation [3]. When number of investigations is concentrated on characterization and modeling of drought, duration, intensity, severity, spatial, and temporal assessment for mapping risk has certain constraints. Liu [4] demonstrated a number of ways to vulnerability mapping with diverse hazards through the conceptual structure, targeting perception level before, during, and after drought onset [5].

This risk is measured as an exposure product to vulnerability and hazard to various conditions. This field is expected to pose higher risk if it shows extreme risk and lower capability for risk impact [6]. Static and climatic events or semi-static parameters like population, technology, practices, behavior, and policies are extremely albeit over longer time scales [7]. This makes vulnerability assessment to extremely challenging factors. Therefore, constant assessment over risk events has to be resolved in existing approaches [8]. With diverse cases in India, these risk events are highly stressful owing to events, there is some necessity for enhancement and appropriate implementation spatiotemporal methods for risk assessments and vulnerabilities [9].

The risk management factor comprises of three essential factors: (a) risk identification and significant assessment; (b) novel method construction and available resource utilization to mitigate and reduce various risks; and (c) novel strategies implementation to handle risk factors [10]. The complexity may enforce any of these factors that are subjective to regional vulnerability measurements that are quantified as relative measures generally [11]. Confronting factors with vulnerability evaluation and mapping are determined to be ongoing issues as vulnerability levels are dynamic, and moderated with variations in land usage, technological factors, population density, farming practices, and variations in climate [12].

Henceforth, mitigating regional factors may include diverse levels of subjective assessments as there are no traditional factors for mapping vulnerabilities to measure vulnerability risk. Moreover, to reduce subjectivity in vulnerability measurement, application of fuzzy logic is encountered in information system for natural mapping in modelling of effectual tools for spatial fuzzy based decision making that embraces Mamdani membership function (MF) to work effectually with number ranging of (0,1) that reflects degree of membership certainty. Indeed of crisp set that may facilitate values on either 1 or 0 as truth levels [13]. The concept behind fuzzy-based function in mapping is to determine spatial functionalities of map as member to set unconstrained fuzzy membership values that relies over (0,1) which ranges with discrete intervals [14]. For diverse complexity crisis like risk management, fuzzy logic is determined to be an essential way to implement and understand, facilitates flexibility by merging various mapping layers that are readily executed with information system and provides spatial objects of diverse measurements into conventional values among 1 or 0.

Fuzzy logic is utilized for flood risk and landslide mapping, evaluating harvesting zones, various hazards evaluation for disaster management, and decision support system for environmental assessment influence between them [15]. Henceforth, fuzzy-based application over diverse environmental factors of risk and hazard management leads to be motivation for utilizing the fuzzy tool for vulnerability computation in various regions of India.

14.2 VULNERABILITY

The vulnerability risk is considered as product or sum of hazards, exposure, and vulnerability. Based on downing evaluation, risk is provided mathematically as in Eqns. (1) and (2):

$$\text{Risk} = \text{hazard} \times \text{vi} = \text{ulnerability} \times \text{Exposure} \qquad (1)$$

$$\text{Risk} = \text{Hazard} + \text{vulnerability} \qquad (2)$$

Various standards are designed for specifying hazard as potential ability of human or nature induced physical event may lead to life loss, injury or other health factors like property loss, infrastructure damage, service provisioning, livelihoods, and diverse environmental factors. Risk is depicted as a livelihood for certain time period of various changes in normal community function or society owing to physical events hazardous that interact with vulnerable conditions. This leads to extensive spread over adverse material, human, environmental factors that need appropriate emergency response to fulfill crucial human requirements and may need external recovery support.

Vulnerability is a term that may possess numerous definitions. As vulnerability have general context-bound, certain factors are complex to justify. In extensive hazard-centric disaster perception, vulnerabilities are provided to an extent which people may suffer from socio-economic and calamities to handle this risk. Some conceptual vulnerability is measured as a hazardous factor that has people, infrastructure, and diverse economics. Finally, some exposure is provided based on resources like 'system, property, people or other elements may provide hazard zones and subjective to potential losses.

14.2.1 MAPPING VULNERABILITY

For mapping vulnerability risk, this work considers cognitive mapping. It is a network node with directed edges, i.e., di-graph which is utilized to provide general relationship among stakeholder's descriptions. This cognitive mapping is considered as a tool which is generally utilized for qualitative modeling of complex systems. Certain essential benefits over these approaches are relatively simple, capacity, and flexible to capture complex model.

Cognitive mapping is anticipated to provide a causal connection between stakeholder's description towards agricultural vulnerabilities. Some practical approaches for the construction of this cognitive mapping is relatively free. Transcription files with computer-based data analysis as cognitive mapping model for evaluating practical systems. Direct mapping with stakeholders during interview, transcription-based mapping for interviews.

For performing agriculture-based coupling and to analyze extreme weather event, cognitive mapping is used to compose variables with perceived influence over eco-system vulnerability in association with extreme weather condition. Variables are associated with one another in relation to causal, oriented, and weighted factor for significant analysis. Outcomes of these steps are computed with cognitive mapping, which is utilized as a primary information source to construct models for evaluation of vulnerabilities to extreme weather events.

14.2.2 FUZZY INFERENCE SYSTEM

The step is considered as an essential phase as it is utilized to provide a model for vulnerability computation in agricultural systems. This approach is based on fuzzy inference. This is used for model construction.

14.2.3 RULE GENERATION

Expertise knowledge towards cognitive mapping is utilized to offer a set of rules that may qualitatively determine influencing key parameters of vulnerable system. Every rule comprises of huge premises and outcomes in a state that $'if(var1 is.) and if (Var2 is...) then(vulnerability is...)$ For all rules: (a) 'var1' and 'var2' is key factors; (b) 'vulnerability' element has to be computed; and (c) '…' are variable levels (for instance, five-point scale: "very high," "high," "medium," "low" and "very low." With two or more premises, it can be merged with Boolean operations like 'or' or 'and.'

14.2.4 FUZZY PARAMETERS

This inference system needs certain technical factors. These values are utilized in this anticipated approach: (a) universe was depicted from $0 \rightarrow 1$ in every 0.01 step; (b) membership functionality with five-point scale using fuzzy with cones of 0.2 universal radius units; (c) fuzzy inference approaches are utilized as general Mamdani approach; (d) every rule outcomes are aggregated with maximal operators; (e) aggregated inference system-based conclusion was defuzzified with centroid approaches as in Figure 14.1.

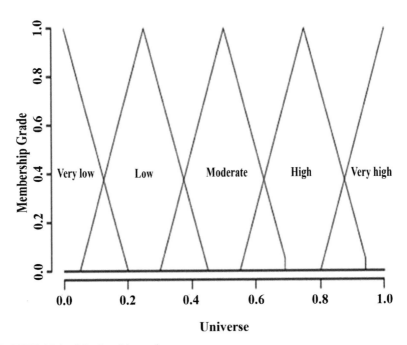

FIGURE 14.1 Membership grade.

14.2.5 INPUT TYPE

Generally, inference system inputs are quantitative values for all key variables. These variables are scaled linearly to handle inference universe among 0 and 1 using 0.01 using Eqn. (3):

$$X_{isc} = X_i / \max(X) \tag{3}$$

A scaled vector value is computed with inference system input. This may return quantitative values over universal among 0 and 1 which specifies assessment towards vulnerabilities.

14.3 NUMERICAL RESULTS

The cognitive mapping and assessment towards agriculture-based system are vulnerability that are provided based on scaled values. The foremost essential fields are positioned in the central locality of countries agricultural region that may specify erosion soil characteristics that are well appropriate for crops and agriculture.

Sensitivity analysis was carried out to compute the impact of every factor over uncertainty of vulnerability assessment. Partial rank correlation coefficients are evaluated with Monte Carlo-based computation (with hypercube size of 50 and 100 bootstrap replications). Outcomes with this sensitivity analysis are provided with land usage significance and soil properties with vulnerability assessment of agriculture to heavy rain. Table 14.1 depicts grade value computation with five scales. Table 14.2 depicts fuzzy set-based risk assessment, and Table 14.3 shows comprehensive model assessment, Table 14.4 is vulnerability computation. Figures 14.2 and 14.3 depicts fuzzy sets and comprehensive grade models. Finally, Figure 14.4 depicts vulnerability computation.

TABLE 14.1 Grade Value Computation

Grade	Risk	Values
1	High	$H < 1.6$
2	High	$1.6 < H < 2.6$
3	Moderate	$2.6 < H < 3.6$
4	Low	$3.6 < H < 4.6$
5	Low	$4.6 \leq H$

TABLE 14.2 Fuzzy Set-Based Risk Assessment

Risk	Fuzzy Set	
	H	Grade
Danger	1.9–2.3	2
Vulnerable	1–1.6	1
Exposure	1.9–2.0	2
Resistibility	3.5–4.0	4
Comprehensive	2–2.5	2

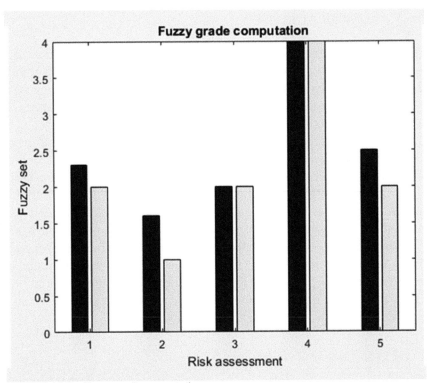

FIGURE 14.2 Fuzzy grade computation.

TABLE 14.3 Comprehensive Model Computation

Risk	Comprehensive Model	
	H	Grade
Danger	1.8	2
Vulnerable	1.6	2
Exposure	2.1	2
Resistibility	4.2	4
Comprehensive	2.3	2

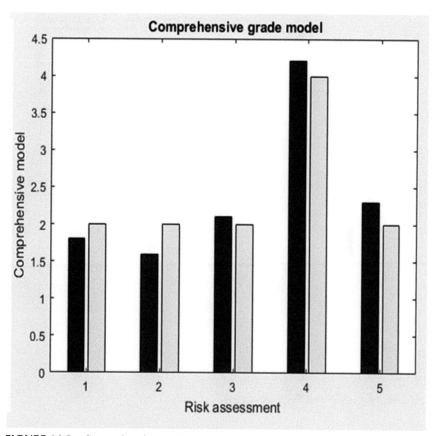

FIGURE 14.3 Comprehensive grade model.

TABLE 14.4 Vulnerability Computation

Vulnerability	Discrete Interval	Area
Very high	0.15–0.50	0.11
High	0.50–0.70	0.23
Moderate	0.68–0.79	0.27
Low	0.79–0.83	0.28
Very low	0.83–0.92	0.16

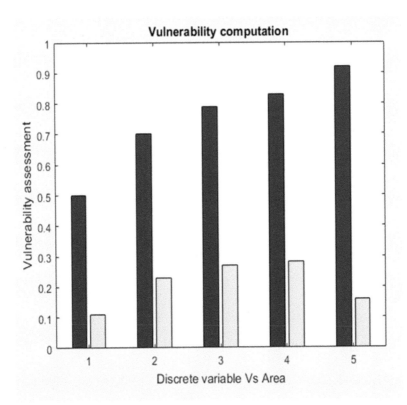

FIGURE 14.4 Vulnerability computation.

The cognitive mapping and assessment towards grassland for analyzing droughts may also change among agricultural fields. Northern parts are extremely vulnerable to drought because of its soil properties, while stocking farms rate which are moderate. Most southern parts are studied with the characterization of higher total available water content.

14.4 CONCLUSION

This investigation uses fuzzy set with inference system and Mamdani MF for vulnerability, dangerousness, drought resistibility and exposure to drought. Based on these assessment-based outcomes, every city may hold lower drought dangerous than western city completely. This may be located with lower drought vulnerability area; however, it may have higher vulnerabilities towards water shortage. Most regions that come under higher drought exposure area is extremely higher. This is because of higher water resources and a lesser proportion of crop planting. Some regions may rank as lower risk exposed city. Some may locate in lower drought resistibility areas and have higher drought resistibility.

Some variable fuzzy sets are modeled here and may perform assessment-based outcomes in interval indeed of valuable points by changing factors. This makes assessment outcomes ore reliable and consistent. Moreover, these universal standards of agricultural risk grade may not prevail over and division methods have to be studied in the future.

KEYWORDS

- climatic variation
- cultivation
- farming
- fuzzy
- management factor
- socio-environmental factors
- vulnerability

REFERENCES

1. Che, S., Chunqiang, L., & Shen, S., (2010). Analysis of drought-flood spatial-temporal characteristics based on standard precipitation index in Hebei province. *Chinese Journal of Agrometeorology, 31*(1), 137–143.
2. Chen, S., (2009). *Theory and Model of Engineering Variable Fuzzy Sets and its Application* (pp. 7–35). Dalian: Dalian University of Technology Press.
3. Huang, Q., Zhang, B., Huang, J., et al., (2007). Research of water-saving society evaluation based on fuzzy matter-element model and coefficient of entropy. *Journal of Hydraulic Engineering, 38*, 413–416.
4. Liu, L., (2007). *Risk Assessment and Risk Management of Agricultural Drought and Flood in Hengyang*. Changsha: Hunan Agriculture University.
5. Wang, S., Yu, C., & He, X., (2008). Assessment of drought and its effect on agriculture in 2006 in Sichuan Basin. *Chinese Journal of Agrometeorology, 29*(1), 115–118.
6. Folke, C., Carpenter, S. R., Walker, B., Scheffer, M., et al., (2010). Resilience thinking: Integrating resilience, adaptability, and transformability. *Ecol. Soc., 15*(4), 20.

7. Gudmundsson, L., & Seneviratne, S. I., (2016). Anthropogenic climate change affects meteorological drought risk in Europe. *Environ. Res. Lett., 11*(4), 044005.
8. Kok, K., (2009). The potential of fuzzy cognitive maps for semi-quantitative scenario development, with an example from Brazil. *Glob. Environ. Change, 19*(1), 122–133.
9. Nelson, R., Kokic, P., Crimp, S., Martin, P., et al., (2010). The vulnerability of Australian rural communities to climate variability and change: Part ii. Integrating impacts with adaptive capacity. *Environ. Sci. Policy, 13*(1), 18–27.
10. Preston, B. L., Yuen, E. J., & Westaway, R. M., (2011). Putting vulnerability to climate change on the map: A review of approaches, benefits, and risks. *Sustain. Sci., 6*(2), 177–202.
11. Dayal, K. S., Deo, R. C., & Apan, A. A., (2017b). Investigating drought duration-severity-intensity characteristics using the standardized precipitation-evapotranspiration index: Case studies in drought-prone Southeast Queensland. *J. Hydrol. Eng., 23*(1), 05017029.
12. Dayal, K., Deo, R., & Apan, A., (2016). Application of hybrid artificial neural network algorithm for the prediction of standardized precipitation index. In: *Proc., IEEE Region 10th International Conference: Technologies for Smart Nation* (pp. 2962–2966). IEEE.
13. Jun, K. S., Chung, E. S., Kim, Y. G., & Kim, Y., (2013). A fuzzy multi-criteria approach to flood risk vulnerability in South Korea by considering climate change impacts. *Expert Syst. Appl., 40*(4), 1003–1013.
14. Liu, K. F., & Lai, J. H., (2009). Decision-support for environmental impact assessment: A hybrid approach using fuzzy logic and fuzzy analytic network process. *Expert Syst. Appl., 36*(3), 5119–5136.
15. Mpelasoka, F., Hennessy, K., Jones, R., & Bates, B., (2008). Comparison of suitable drought indices for climate change impacts assessment over Australia towards resource management. *Int. J. Climatol., 28*(10), 1283–1292.

CHAPTER 15

MODELING AN INTELLIGENT SYSTEM FOR HEALTH CARE MANAGEMENT

A. JAYANTHILADEVI,[1] P. S. AITHAL,[2] K. KRISHNA PRASAD,[1] and MANIVEL KANDASAMY[3]

[1]Computer Science and Information Science, Srinivas University, Mangalore, Karnataka, India, E-mail: drjayanthila@gmail.com (A. Jayanthiladevi)

[2]Srinivas University, Mangalore, Karnataka, India

[3]Optum Tech-United Health Group, Bangalore, Karnataka, India

ABSTRACT

The ultimate objective of modeling this intellectual system is to provide appropriate information regarding the emergency condition. The significant objective of this work is to model an intelligent system for computing the assistance provided by hospitals during emergency condition. The model implementation assists in providing an effectual approach to enhance waiting time process and to improve health care quality. This model is more effectual, and it is based on patients' health conditions and user's decisions.

15.1 INTRODUCTION

Recently, the need towards emergency health care conditions is increased, and thus intellectual management systems are more essential [1]. To be specific, the foremost issue is to handle the total number of patients in daily basis. This turns to be serious issues all over the world as it needs the usage of vast resources both in materialization and human needs that are extremely restricted [2]. It is stressed that coordination degree among these situations is increased number of patients with saturation condition not fulfilled.

The outcome is the increased amount of time to be given during services from the initial stage to the final stage. This is known as staying time. The most prominent parameters are to provide effectual QoS. When this QoS is not fulfilled [3], a saturation point is triggered, and some discontentment is identified between them. This includes lesser access to an emergency condition, increased patient mortality, and reduced care. The most complex factor encountered in a hospital environment is owing to the lack of dynamicity and variability of healthcare devices [4]. The services are considered to be the outcome of the interaction between the available elements. The simulation and modeling of these systems are considered to be essential tools for characterization. This assists in providing appropriate decision-making to setting an optimal system for handling complexities.

The ultimate objective of modeling this intellectual system is to provide appropriate information regarding the emergency condition. It is probable to attain an inference process with system variables to make prediction based on variable characterization [5]. In various studies, emergency condition-based simulation model is designed. This is validated with the research community by collaboration with various hospitals. This model provides the basic characterization of services with agent interactions. This is modeled with a multi-agent programming environment for complex circumstances. With non-critical patient's record, some patients do not need huge attention and higher patient's treatment environment. This state is also considered as saturation point to patient admitted. Some scheduling models are anticipated at the initial stage of patterns over non-critical conditions [6]. This is determined to be an entry-level with patient's history and relocation is done with the enhancement in waiting time that influences the patients. Thus, enhanced quality of services is needed.

Some simulators are considered as an evaluation tool and outcomes the development of the anticipated model. The simulation is to provide data when needed and to deal with various circumstances by model applications [7]. The scheduling point and health care configurations are needed, and it is not feasible to validate the approaches in various scenarios, and the cause is to acquire essential information for validation purposes. The essential contribution of this work is to develop an intellectual approach for evaluating the throughput of the system model. This model determines the ability to deal with the available patients, and the goal of modeling is to compute responsive ability based on available staff configurations [8]. This analytical model deals with a number of personal details, experience level, and care provided during the treatment process. This is analyzed and validated based on records generated during real-time simulation model. Figure 15.1 depicts the generic view of health care monitoring system.

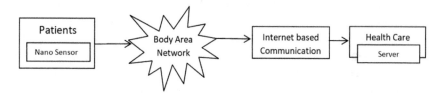

FIGURE 15.1 Generic view of health care monitoring system.

The subsequent is to provide an appropriate definition for model scheduling for some non-critical patients and to give better services. The flow pattern is supplied by hospitals.' This offers a way to compute responsiveness for staff over healthcare patients. This is validated by effectual schedule modeling with the intellectual devices and to record the data for computing the results [9]. This contribution towards the intellectual management model is based on service behavior and to provide appropriate decision-making conditions for enhancing QoS and demands service growth in the future.

15.2 RELATED WORKS

Conventional behavior interventions the access towards health care monitoring. Some real-time monitoring tools are considered to be more feasible, and some actions are taken over a certain time period. Integrated wearable devices and sensors provide higher-level data generation to sensors. The process of data transformation like running, walking, and standing are more recursive and thus constructs a system model [10]. The data can be accessed, and some activities are triggered. The wearable devices are considered to be more personalized and attractive. It fits with daily activities [11]. It is probable to utilize audio to identify the stress level computation and mood transitions. This is extremely valuable for dealing with behavioral patterns like depression and bipolar disorders. This helps to predict the functionalities like partying. The health care goals are predicted automatically. This is reported manually and makes it easier for long time usage. Digital health care systems like E-mail, calendars, and social networks are also used.

Digital health monitoring system shows interventions of psychological and behavioral characteristics to assist mental and physical health. The solution is to use digital media to targeted characteristics, emotions, and cognitions. This comprises of psychological and behavioral strategies towards targeted mental and physical health [12]. This may offer promising solutions to have psychological factors. This is triggered based on the efficiency of the device to predict diet, depression, anxiety. The complete potential of this model is used to validate the psychological activities in technological modeling. As an outcome, this is used for proof-based interventions [13]. In contrary to face-to-face interventions, health monitoring system is non-consumable. It is utilized by various patients to transcend geographical constraints and diminishes the disparities. It is utilized simultaneously and integrated with daily functionalities with reinforcement, unobtrusiveness, pleasantness to offer acceptable privacy levels. It is suggested to compute the usage of present technological level and to measure the characteristics of principle translating level from other modalities [14]. The notification is generated during the matching period. When probability is higher, users have the opportunity to read the notification and reads the content and initiates the interactions.

Sometimes, the user may discard or ignores the notification, and reception is considered to be very lesser. Here, some factors are receptive and not predicted at the moment. The user reaction is an interruption and subjective to experience it very often. The matching time is determined and notifications are generated. Mobiles are used as notification devices and carry out certain functionalities. It is advantageous to some users, notification, and demand attention of certain moments. Users quickly respond towards notification when seems to be more idle. Sometimes, users may be annoyed with notification content [15]. The response is more interesting, appropriate, active, and entertaining. The communication is determined to be more essential. Time criticality and pressure are more appropriate. Some mechanisms are favorable to notification. The response time and disruption is influenced by alert, sender-recipient relationship and task complexity is engaged. Some notification shows information and disruption. This may adversely influence task completion time, error rate and emotion and user affection. With interference uses various machine learning (ML) classifiers to include activities, location, and notification content.

15.3 METHODOLOGY

The intelligent management system comprises of some hardware component. The patients are provided with wearable devices and utilized to register and to offer feedback and intervention. Here, smartwatches are utilized. Some sensors are required to track patients' functionality. It is utilized to cover various regions to improve accuracy of activation tracking. The sensors track the functionality of patients' concurrently. The tracking and facial gestures are used for tracking system. The computers are used for tracking functionality. It is connected with USB cables and wirelessly. It is utilized for computing the kinetic motion of sensors and to track the users' concurrently. The phone is utilized to bridge communication among computer and consented worker. Phone is used to communicate controlling factors with Wi-Fi. This is permanently used to close track with corresponding users.

Some software components are used for implementation and tied with certain wearable technologies that show a lack in wearable communication. The software components are used, and higher-level software is used to interact with one another. This provides an interface among the register and worker to acquire alert notifications and to show performance and application configuration. It receives motion tracking data from connected system. The users can interact with the core system and includes registration, tracking, logging, and alert generation. This works as servers and application towards the client. The component is utilized to establish communication among the tracking system and application through standardized HTTP.

The patient's location is tracked and achieved with the use of computer vision (CV)-based tracking and registration mechanism. It introduces some mechanism for the identification of users and does not register the system usage manually. In case of registration, the users have to wear the device and to register it with the system; the connection is established among the devices and phone. This predicts the unique ID and known as user id. It safeguards and records the user id. The data is connected with the system to measure the complete identification. The hospitals have to store the information more safely during offline condition also. When entering the hospital environment, it is equipped with a tracking system, and the user initiates the registration process and to validate the tracking applications. This may fulfill reliability. It involves two steps, pressing button and non-natural gestures. Some physical button is given to the user for pressing during emergency condition. The gesture verification is used for tracking application. This is utilized with certain configuration setup and gesture is easier. It reduces the error during registration process. It can be used in various other mechanisms. The registration process is validated with diverse motion tracking process through static rules. This represents the angle of the body position and shows a higher threshold to ensure the position. The parameters require the tuning process and the experiments are validated with certain device utilization.

Once the process is completed, the identification and registration process initiates the tracking system and users' have to concentrate on the system. The registration process is used for predicting the patients and shows its usability towards the patients in need of emergency conditions. This model is utilized for reducing the complexity of the users and improves system usability. The user has to register the biometrics more sufficiently. The patients explicitly register with the system and the records are noted.

The data is stored in memory for reducing the fault and works to eradicate fault tolerance. It is compared with logged biometric data. When the match is found, the patient is identified automatically. The features are more important for validating the differences. The vectors are more prominent across various sessions and orientations are provided to it. The patients' condition is analyzed with various experimentation, when patients logged into the system. To show the distinctiveness towards the feature vector, the segment length and dimensions are measured. The length and segments are merged for computation. The 2D vectors are provided for separation. For various pairs of users, it provides lesser distance among the pair of users while comparing with the other feature vectors. When the threshold is chosen, the vector leads to a set of users. When the value is lesser below threshold, it cannot be differentiated from the others. The dimensionality factor reduces the computational needs of the system. While providing the process flow, threshold for various users are given. The parameter is used for computation of average distance among the patients for predicting and registering the workers. To demonstrate the approximation, two factors are considered:

1. Inter-user distances are distributed, and differences among the users are validated with threshold value. The values must be greater than the other form of distances and to give improved classification rate; and
2. The distance distribution and tracking errors are discussed. The vector values are fluctuated and the number of frames is measured. The threshold value is directly influenced by the window size. When the value is smaller, a larger window size is considered. It sometimes leads to delay in predicting the patients.' The prediction speed should be matched with the user frame. To have a balance among the matching threshold and window size, certain factors should be measured.

Some feasibility measures should be considered and relatively connected with the differences among various conditions: (a) fixed orientation; (b) similar orientation; and (c) diverse orientation. The computation among the frames is measured with the feature vectors. The data is collected from the frames and to compare it with available devices for logging the patient details. It is not so reliable and utilizes a smaller threshold as it does not misclassify the user data due to variations. Therefore, a balanced model has to be designed with feature vectors and window size. Next, user consistency has to be determined. It is evaluated by comparing the distance among the mean vector value with the overall vector value. The outcomes are measured, and the trails are observed. The distance may vary from 0.0010 and 0.038 m which is extremely below the threshold value which is set more heuristically.

When the patient enters the view, patient data is taken and compared with biometric data. There is an exhaustive search over the logged data as it is held only when the patient wears the device. When the patient is under the view, the connection is established automatically and user id is generated to monitor the tracking activities. This is sometimes restricted and user-id ranges from motion tracking functionality. The decision rules are generated when the patients stay nearer to the load. The rules are lesser needed when angles exceed the pre-defined threshold value. The load is measured based on patients' condition. The invariance determines the angle and position of patients. The configurations are provided based on the vector values. The invariance rule is complex to be measured and expressed based on system configuration. The boundaries and configurations are more needed. When the condition is violated and it exceeds the threshold value, the domain knowledge is not fulfilled. The intelligent management system is not aware of the connection establishment and the process is declined.

15.4 NUMERICAL RESULTS

The simulation is performed in MATLAB environment. Here, various experiments are carried out to shows the functionality of the intellectual system. The patients stand before the sensor devices and face it with an approximate angle of 45°, the right-side view is also 45°. To compute the overall performance, various participants are included to perform the experiment. The device position is crucial to validate the system performance.

The system is extremely robust and the lower extreme is more occlusion. This is done while rolling the patients to opposite direction of devices. It is obvious that some activities are more common and un-tracked with un-registration process. The process is repeated and manual registration is carried out initially. The entry details are depicted in Table 15.1.

TABLE 15.1 Entry Details

Time	1	2	3	4	5	6	7	8	9	10
Entry Details	4	3	2	5	1	2	4	9	10	6
Movement	1	1	1	0	0	0	0	1	2	6
Care Given	5	4	3	2	1	2	2	4	8	13

Some limitations are also identified. Based on the position, the scheduling process is determined. The location of the patient is monitored in hourly basis and identifies the critical hours. Based on the location range, the matching process is initiated in critical hour for eliminating the holes in treating patients. The position of the user is re-computed under various circumstances. In the initial process, the algorithm updates and generates the schedule process of the patients. Later, the patients are provided with an appointment and relocated based on arrival time. The creation and characterization is based on total available patients.

The patient relocation process is performed to improve the system attention. The index value for the scheduling process is measured, and the constraints are validated. The outcomes are based on system ability. The relocation is based on the patient movement and reduces the waiting time to offer QoS. The non-critical patients are relocated, and maximal patients attended without waiting time is also computed. The maximal delay time should reach from ½ hour to 1 hour. It determines the critical hole and number of surpass. Based on relocation, it eliminates the tentative patients and generates hole when possible. The propagation is re-computed. Figures 15.2 and 15.3 measures the patient's details for performing scheduling function.

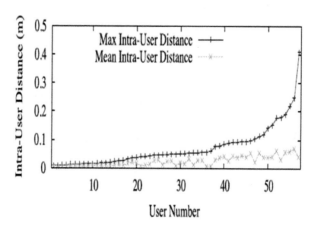

FIGURE 15.2 Patients' distance computation for scheduling the arrival time.

The updates are generated periodically and specify the admittance hour of patients. The appointment for relocated patients is based on arrival time. Various new functionalities are constructed by the intelligent system. This is based on entry parameters and the updation process. The recommendation table for relocation is provided when needed. The assignment process is more essential and done with first come first serve method. The distance is measured among the patients as depicted in Figure 15.4.

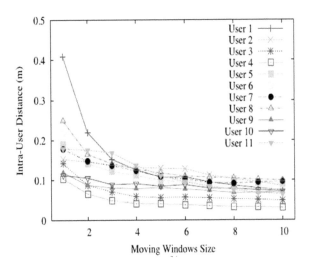

FIGURE 15.3 Window size based on patients' movement.

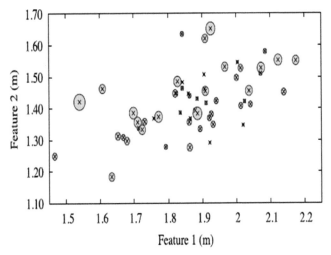

FIGURE 15.4 Distance measurement.

15.5 CONCLUSION

The significant objective of this work is to model an intelligent system for computing the assistance provided by hospitals during emergency condition. This is achieved by matching process, scheduling process and vector values. The scheduling idea is to improve the quality of services offered by the physicians, and some essential tools are given for service administration and to measure the growth of various services. The model implementation assists in providing an effectual approach to enhance waiting time process and to improve health care quality. This model is more effectual, and it is based on patients' health conditions and user's decisions.

KEYWORDS

- **decision making**
- **emergency condition**
- **health care management**
- **intelligent system**
- **materialization**
- **waiting time**

REFERENCES

1. Indrakumari, R., Poongodi, T., Suresh, P., & Balamurugan, B., (2020). The growing role of internet of things in healthcare wearables. In: *Emergence of Pharmaceutical Industry Growth with Industrial IoT Approach* (pp. 163–194). Amsterdam, The Netherlands: Elsevier.
2. Kang, S., (2020). A study on smart homecare for daily living ability and safety management of the elderly. In: *Information Science and Applications* (pp. 707–710). Singapore: Springer.
3. Hassan, M. M., Alam, M. G. R., Uddin, M. Z., Huda, S., Almogren, A., & Fortino, G., (2019). Human emotion recognition using deep belief network architecture. *Inf. Fusion, 51*, 10–18.
4. Puntambekar, V., Agarwal, S., & Mahalakshmi, P., (2020). Dynamic monitoring of health using smart health band. In: *Soft Computer for Problem Solving* (pp. 453–462). Singapore: Springer.
5. Ismail, W., & Hassan, M., (2017). Mining productive-associated periodic frequent patterns in body sensor data for smart home care. *Sensors, 17*(5), 952.
6. Sahoo, A. K., Pradhan, C., Barik, R. K., & Dubey, H., (2019). DeepReco: Deep learning-based health recommender system using collaborative filtering. *Computation, 7*(2), 25.

7. Ismail, W., & Hassan, M., (2017). Mining productive-associated periodic frequent patterns in body sensor data for smart home care. *Sensors, 17*(5), 952.
8. Malvoni, M., De Giorgi, M. G., & Congedo, P. M., (2016). Data on support vector machines (SVM) model to forecast photovoltaic power. *Data Brief, 9,* 13–16.
9. Ismail, A., Shehab, A., & El-Henawy, I., (2019). Healthcare analysis in smart big data analytics: Reviews, challenges, and recommendations. In: *Security in Smart Cities: Models, Applications, and Challenges* (pp. 27–45). Cham, Switzerland: Springer.
10. Pham, T., Tran, T., Phung, D., & Venkatesh, S., (2017). Predicting healthcare trajectories from medical records: A deep learning approach. *J. Biomed. Informat., 69,* 218–229.
11. Liu, Y., Zhang, Q., Zhao, G., Qu, Z., Liu, G., Liu, Z., & An, Y., (2019). Detecting diseases by human-physiological-parameter-based deep learning. *IEEE Access, 7,* 22002–22010.
12. Miotto, R., Li, L., Kidd, B. A., & Dudley, J. T., (2016). Deep patient: An unsupervised representation to predict the future of patients from the electronic health records. *Sci. Rep., 6*(1). Art. no. 26094.
13. Ebadollahi, S., Sun, J., Gotz, D., Hu, J., Sow, D., & Neti, C., (2010). Predicting patient's trajectory of physiological data using temporal trends in similar patients: A system for near-term prognostics. In: *Proc. AMIA Annu. Symp.* (p. 192).
14. Kweon, S., Kim, Y., Jang, M. J., Kim, Y., Kim, K., Choi, S., Chun, C., Khang, Y. H., & Oh, K., (2014). Data resource profile: The Korea national health and nutrition examination survey (KNHANES). *Int. J. Epidemiol., 43*(1), 69–77.
15. Yoo, H., & Chung, K., (2018). Heart rate variability-based stress index service model using bio-sensor. *Cluster Comput., 21*(1), 1139–1149.

INDEX

A

Accumulated median error (AME), 34, 35
Activation
  function, 143, 156, 157, 159–164, 166, 174, 189, 200
    rectified linear unit (RELU), 162
    Saha Bora activation function (SBAF), 164, 165, 174
    sigmoid activation function, 159
    Softmax activation function, 163
    tanh activation function, 161
  tracking, 241
AdaBoost, 170
Adaptive
  cruise control, 3
  invasion-based model (AIM), 30, 31, 37
Advance
  driving assistance system, 3
  encryption standard (AES), 72, 84, 85, 87
Aggregated fuzzy ratings (AFR), 98, 101
Algorithm
  bias, 4
  dependent parameters, 24
  implementation, 7
Ant
  colony optimization (ACO), 1, 2, 7, 8, 10, 12–15, 17, 19–21, 24, 37, 121, 125–127, 129, 130, 132
    algorithm, 2, 7, 10, 12, 14, 129, 130
    ants in nature, 8
    imagery edges detection, 16
    metaheuristic, 12
    real ants vs. artificial ants, 10
    solve travelling salesman problem (TSP), 14
  system (AS), 14, 130, 131
Antagonistic language, 61
Anti-fuse elements, 74
Approximation, 24, 200, 243
Arithmetic
  and logic unit (ALU), 70, 71, 76, 78, 87
  operations, 71, 76
Array multiplier, 77
Artificial
  agent, 8
  ants, 7, 10, 11, 20
  applications, 169
    agriculture, 172
    astronomical objects classification, 174
    character recognition, 169
    healthcare, 171
    image processing, 169
    loan application evaluation, 171
    self-driving cars, 172
    stock price prediction, 170
  architecture, 155
  bee colony (ABC), 7, 28, 33–37
    optimization (ABCO), 7
  environment, 11
  errors, 105
  immune system (AIS), 24, 30, 37
  intelligence (AI), 1–5, 12, 18–21, 64, 122, 124, 125, 179, 181
  networks (ANNs), 153, 158, 171, 174, 177, 178, 180, 184, 185, 187, 193
  neural computing (ANC), 24, 37
  pheromone, 7
  systems, 1, 125
  vision, 23
Asymmetric
  key encryption, 84
  traveling salesperson problem (ATSP), 130
Authorship identification, 177–180, 184, 193
Automated detection, 62
Automatic
  pruning, 190
  speech recognition, 59, 62
Axons, 154

B

Backpropagation, 153, 157, 159, 164, 166, 180, 200
Bacteria molding, 5
Bacterial foraging optimization, 2, 7
Basic declivity edge detector, 17
BAT algorithm, 127
Bayesian
  approach, 204
  classifier, 198
  inference, 198
  rules, 201
Bee colony optimization (BCO), 2, 7
Bifurcation, 41, 42, 44, 51, 52
  analysis, 51
  diagram, 41, 42, 44, 51, 52
Big data, 64, 195, 197, 204
Bigrams, 140
Bilingual evaluation understudy (BLEU), 136, 140, 141, 145, 146, 148, 149
Binary
  amplitude shift keying (BASK), 71, 87
  classification model, 159
  decision, 129
  frequency-shift keying (BFSK), 71, 81
  optimization problem, 137
  phase-shift keying (BPSK), 71, 72, 81, 83, 84
  PSO (BPSO), 2, 7, 8, 24, 32, 125–129, 131, 132, 217, 218
Biogeography-based optimization, 2, 7
Biography-based optimization (BBO), 2, 7, 28, 33–37
Bio-inspired algorithms, 2, 19
Biological
  invasion, 31
  neural network architecture, 154
Biometric data, 242, 243
Bitstream generation, 85
Black box, 4, 83
Boolean
  function, 17
  operations, 229

C

Candidate
  solution, 12, 31, 37
  pool (CSP), 31, 32, 37
  summary, 145, 146, 148
Canny edge detector, 17
Cardiovascular
  drug, 171
  medicine, 171
Case-specific learning, 4
Cat swarm optimization (CSO), 7, 29
Cellular robotic systems, 20
Chaotic system, 41–44, 46, 49, 51, 55, 56, 57
  qualitative analysis, 44
    asymmetry, 44
    dissipativitiy, 44
    solution existence and uniqueness, 46
Climatic variation, 235
Cloud, 25, 121–126, 132, 208, 217
  based environment, 132
  center, 122
  computing, 25, 121, 122, 124–126, 208
  data center, 121, 122
  environment, 123, 125
  infrastructure, 126
  providers, 122, 125
  service models, 122
  traffic engineering, 124
Clustering, 8, 169
Cognitive mapping, 228, 229, 231, 234
Coherent global pattern, 5
Combinatorial optimization, 12
Common bag of words (CBOW), 64, 66, 138, 139, 191, 192, 191, 193
Communication
  block, 72
  medium, 12
Competitive algorithm, 131
Computational
  comparison, 23
  intelligence, 125, 210
  memory, 122
  problems, 13, 19, 129
  study, 25, 26, 28, 36
Computer
  aided design, 23
  paradigm, 122
  science, 23, 59, 60
  vision (CV), 20, 23, 25, 36, 155, 172, 241
Configurable logic block (CLB), 72–74, 87
Configuration memory, 71, 75

Index

Confusion matrix, 146–148, 189, 190
Construction-step index, 13
Control parameters, 23, 24, 30, 31, 33, 36
Conventional
  (hard) computing, 24
  algorithms, 123
  data analytics, 198
  design techniques, 70, 75
  fertilization, 211
  tools, 214
Convolutional neural network (CNN), 135, 136, 138–140, 142, 143, 148, 149, 155, 162, 169, 170, 172, 174, 201, 202
Co-occurrence matrix, 187
Coronary artery disease, 171
Cortex, 154
Cost-effective tool, 171
Credibility
  measure (CM), 105, 106, 108, 110, 111, 117
  theory (CT), 29, 105, 106, 117
Credit risk assessment, 13, 14
Crime pattern theory, 64
Critical decisions, 4
Cross-entropy, 143, 174
Cuckoo search, 2
Cultivation, 210, 217, 218, 221, 235
Cutting-edge population-based optimization algorithms, 28
Cyberbullying, 62
Cyberhate, 61

D

Data
  alignment, 201
  centers, 121, 122, 125, 126
  communication, 72
  computation systems, 71
  encryption algorithm, 72
  enter applications, 125
  extraction, 20, 201
  governance, 198
  pertaining, 84
  privacy, 4
  processing, 170, 178, 197, 222
  sampling, 5
  scarcity, 5
  storage, 5
Dataset, 5, 25, 30–37, 61, 65, 130, 141, 143, 145–148, 156, 166–169, 178, 184, 187, 188, 197, 200
Decision
  making, 3, 89, 90, 97, 101, 105, 196, 209–211, 227, 238, 239, 247
  support system, 196, 227
Deep learning, 62–64, 66, 178
Degree of overlapping, 26
Dendrites, 154
Dense group, 5
Depth discontinuity, 17
Design rules check (DRC), 76, 87
Differential evolution (DE), 28–31
Digital
  computing systems, 71
  health monitoring system, 240
  healthcare models, 196
  metadata, 64
  modulation, 81
  system design, 69, 70
Dimensionality reduction, 222
Discrete PSO algorithm (DPSO), 129
Dislocated phase synchronization, 57
Distributed
  bag of words version of paragraph vector (PV-DBOW), 138
  DE (dDE), 30, 31
  memory
    model, 181
    version of paragraph vector (PV-DM), 138
DNA sequencing, 12
Doc2Vec, 138, 139, 142, 148, 149, 177, 179–181, 186, 187, 192
Double
  compound combination, 41
  scroll attractor, 42
DriveNet, 173
Drought resistibility, 226, 234, 235
Dynamic, 42, 44, 49, 51, 55
  reconfiguration, 84
  resource allocation, 125
Dynamical analysis, 57
Dynamicity, 238

E

Echolocation, 127
Edge, 2, 14, 17, 156, 166, 169, 178, 199, 228
  detection, 14, 16, 17, 20, 21, 140
Eigenvalues, 48, 49
Electrocardiography, 171
Electronic media, 60
E-mail, 3, 84
Embeddings, 138, 149, 177–180, 183, 184, 189, 191–193
Emergency condition, 237, 238, 242, 247
Emotional robots, 4
Empirical analysis, 64
Encryption algorithms, 70–72, 84, 85
Equilibrium, 41, 42, 44, 47–49, 52, 56
  point, 42, 47, 48, 56
    analysis, 41, 44, 52
Equivalent vector, 142
Error
  gradient, 157, 158, 164
  rate, 160–162, 166, 217, 241
Evolution strategies (ESs), 24, 32
Evolutionary
  algorithms (EAs), 23–25, 30, 36
  computation (EC), 3, 24, 25, 27, 28, 34
    algorithms, 24
  programming (EP), 24
  rigid-body docking (ERBD), 29, 30
Expected value (EV), 105, 106, 112–115, 117
Exploitation, 8
Exploration, 7, 8, 187, 198
Extractive summarization, 136, 149

F

Facebook, 3, 170, 179, 181
Facial recognition, 155
Factual data, 143
False solutions, 31
Farming, 211, 227, 235
FastText, 65, 177, 179, 181, 182, 190–192
Fault
  free conditions, 76
  tolerant model, 159
Feature learning, 139, 140
Field
  programmable gate arrays (FPGAs), 69–75, 87
  specialist scarcity, 5
Firefly
  algorithm, 2
  optimization (FFO), 32
Fish schooling, 5, 7, 24
Fitness function, 8, 12
Flaming, 62
Flocking, 5, 7, 24, 127
Flower pollination by artificial bees (FPAB), 2, 7, 8
Food source, 9, 10, 12, 28
Foraging, 2, 5, 7, 9, 10, 129
Forward propagation, 157, 165, 166
Founding subpopulation, 31
Frequency, 62, 63, 71, 137, 146, 187
F-score, 63
Fuzzy, 24, 89, 90, 97, 98, 101, 105–107, 115, 118, 207, 209–222, 227, 229, 231, 232, 234, 235
  based recommendation system (FRS), 214, 222
  inference system, 229
  model, 213, 214, 216, 217, 222
  parameters, 229
  partitions, 215
  rules, 214
  set theory (FST), 89, 90, 101, 105, 118
  systems (FS), 24, 90
  variable (FV), 105–107, 112–115, 118, 215

G

Gabor wavelets, 169
Gamma, 174
Gaussian mixture model, 137
Generate bit stream, 76
Genetic algorithm (GA), 24, 27, 126, 137, 170, 214
Geometric
  /parametric function, 26
  mean, 141
Germination phase, 210
Global
  optimal solution, 32, 128

optimization, 27
optimum, 29
reconfiguration, 74
statistical information, 183, 187
vector (GloVe), 177, 179, 183, 184, 187–189, 191–193
representation, 183
Glowworm swarm optimization, 7
Gradient descent, 166–169, 174, 180
Gravitational search algorithm (GSA), 125, 126, 132
Green computing, 122, 124, 126

H

Hardware resources, 69, 70, 122
Harmony search (HS), 28, 33–37
Hate speech, 59, 60, 62–67
Hazard-centric disaster perception, 228
Healthcare
  data, 195–197
  management, 247
  monitoring system, 239
Heuristic
  algorithms, 32
  information matrix, 17
  matrix, 13, 16
  methods, 124
Hierarchical cluster analysis, 90
Hybrid
  systems, 62
  technique, 126
Hybridization, 28, 32, 212
Hyperchaotic systems, 42

I

Illumination discontinuity, 17
Image
  acquisition, 16
  compression, 16
  encryption, 41, 55, 57
  enhancement, 16
  processing, 1, 2, 13, 14, 16, 20, 21, 25, 169
  registration (IR), 23, 25–32, 36
    algorithm, 29
    method, 25, 26, 28, 31, 32

optimization procedure, 26
problems, 25, 29–32
registration transformation, 26
similarity metric function, 26, 30
solutions, 26, 30
transformation, 26, 29
restoration, 16
segmentation, 16, 20, 169
Implementation phase, 73
Importance weight of criteria (IWC), 97, 99
Imprecision, 24, 214
Individual
  modulators, 72
  simplicity, 18
Industrial control systems, 25
Information technology, 208, 211, 222, 225
Infrared, 174
Infrastructure as a service (IaaS), 122
Initial
  conditions (I.C.), 41, 43, 47, 49, 51, 52, 55
  fuzzy partition, 215
Inorganic fertilizers, 210
In-phase component, 81
Input matrix, 140
Integer solutions, 129
Integrated logic analyzer (ILA), 84, 85, 87
Integration challenges, 5
Intelligent
  machines, 3
  optimization, 127
  system, 14, 237, 245–247
Interaction energy, 30
Interconnection mechanism, 73
Inter-disciplinary factors, 225, 226
Intergroup theory, 61
Internal configuration access port (ICAP), 72, 75, 84, 85
Intuitionistic fuzzy set (IFS), 89, 90
Inverse credibility distribution (ICD), 107, 112–114, 117, 118

J

Jaccard index, 89
Jacobian matrix, 48, 49

K

Kaplan-Yorke dimension, 42, 44
Konica Minolta c laser range scanner, 30

L

LaneNet, 173
Latent Dirichlet association, 178
League championship algorithm (LCA), 127, 132
Learning model, 195, 196, 201, 204
LightNet, 173
Linear
    matrix, 180
    unit, 143, 149
Linguistic
    rating variables (LRVs), 97, 101
    variable, 99, 100
    weighting variables (LWVs), 97, 99
Liouville's theorem, 45
Local
    optima solutions, 26
    subpopulation, 31
LoG edge detector, 17
Logic
    cells, 73, 74
    gates, 77
Long short-term memory (LSTM), 64, 66, 67, 155, 170
Lorenz system, 42
Loss function, 65, 143, 174
Lyapunov exponent, 41, 50, 57

M

Machine
    generated summary, 144, 145
    learning (ML), 3, 4, 59, 60, 62, 64–67, 155, 171, 172, 178, 179, 196, 204, 241
    algorithms, 59, 172
    techniques, 66, 171, 179
    migration techniques, 126
    vision, 20
Makespan, 125
Mamdani approach, 229
Management factor, 226, 235
MapNet, 173
MapReduce, 204
Marr-Hildreth edge detector, 17
Master system, 52
Matching-based approach, 26
Materialization, 238, 247
Matrix notation, 44
Max pooling, 140, 142
Mean-square-error, 157
Medical imaging, 23, 26
Membership function (MF), 90, 91, 106, 107, 114, 118, 211, 213, 214, 216–220, 227, 229, 234
Memetic algorithm (MA), 30, 218, 221
Memory
    blocks, 72
    elements, 72
    slots, 11
Message passing interface (MPI), 31
Meta heuristic-based techniques, 126
Metadata, 64, 65
Metaheuristic, 12, 20, 21, 24, 127
    approach, 7
    methods, 122, 124, 127
Microprocessors, 71
Minion investigators, 105
Modulation functional blocks, 70
Momentum, 62
Monte Carlo-based computation, 231
Multilayer perceptron (MLP), 155, 170, 174
Multi-objective PSO (MOPSO), 129
Multiple algorithms, 72
Multi-switching phase synchronization, 41, 42, 52, 55, 56
    dislocated phase synchronization, 53

N

Natural
    intelligence, 3
    language processing, 3, 155, 178
    selection, 8
Nature-inspired
    algorithms, 10, 121, 123
    computing, 125
Necessity measure (NM), 106, 108, 110, 117, 118

Index   255

Negative
　definite, 55
　real part, 48, 49
Network
　architecture, 125, 148
　layers, 18
　topology information, 127
Neural
　embeddings, 178, 179, 184, 193
　language, 63, 64
　network, 3, 62, 66, 67, 135–139, 142, 148, 149, 153–156, 158, 159, 164, 166, 168, 169, 171, 172, 174, 177–180, 190, 193, 195, 200, 204, 214
　　applications, 174
Neuron, 6, 153–159, 162, 164–166, 174, 178, 200
N-grams, 64, 140, 179, 181, 182
Non-conventional algorithms, 123
Non-integer value, 50
Non-linear
　control functions, 54, 57
　functional layers, 196
　optimization problem, 24, 121
Non-natural gestures, 242
Non-numerical data, 178
Non-zero real part, 48, 49
Novel
　chaotic system, 43, 57
　scheme, 125
　synchronization technique, 41, 42, 55
　system insecure communication, 55

O

One time programmable (OTP), 73, 74
Online
　media, 63
　social networks, 61
OpenRoadNet, 173
Optimal
　configuration, 29
　resource utilization, 122, 125
　solution, 14, 126, 128, 129, 132
　transformation, 25
Optimization
　algorithm, 24, 25, 32, 121, 123, 124, 127, 166

　batch gradient descent, 167
　mini-batch gradient descent, 167
　stochastic gradient descent, 168
　problems, 8, 12, 17, 20, 121, 129, 130
Orientation
　diverse orientation, 243
　fixed orientation, 243
　similar orientation, 243
　surface orientation, 17
Orthogonal
　crossover, 29
　learning (OL), 28, 29
Othering, 60–62, 66
Output layers, 156, 165

P

Paragraph2vec, 63, 64
Parameter-based approach, 26
Parametric ReLU (PReLU), 174
Partial
　blocks (p-blocks), 76
　rank correlation coefficients, 231
　reconfiguration (PR), 69–72, 74–81, 83–87
　　design methodology, 74
　truth, 24
Particle swarm optimization (PSO), 2, 24, 121, 125, 127, 132
　variants, 129
Path planning, 2, 14, 20, 21
PathFinders, 173
PathNet, 173
Patient-centric facilities, 171
Pattern recognition, 155
P-block, 76, 79
Perceptron, 153, 154, 174
Perturbed
　global best (p-gbest), 128, 129
　PSO (PPSO), 129
pH, 210
Pheromone, 7–9, 11–13, 16, 17, 130, 131
　evaporation, 11
　matrix, 13, 16, 17
　path, 7
　trail, 7, 11, 12
Photometric observations, 174
Picture fuzzy set (PFS), 89–91, 93
Plagiarism, 178

Platform as a service (PaaS), 122
Poincare
  map, 42
  surface, 51
Policymakers, 225, 226
Pollen, 8
Pollination, 2, 8
Polynomial time, 24, 126
Pooling, 140
  layers, 139, 174
Population-based approach, 12, 28
Positive-negative keywords, 137
Possibility measure (PM), 106, 108, 109, 117, 129
Predicaments, 1, 11
Preliminaries, 90–92, 106
  fuzzy set (FS), 90
  IFS, 90
  PFS, 90
    alpha cut, 91
    trapezoidal PFS (TRPFS), 92
    triangular PFSS (TPFS), 91
Priority aware VM allocation (PAVA), 127
Pro-active risk management, 226
Probabilistic transition matrix, 13
Processor section (PS), 85
Projection matrix, 200
Proposed hybrid approach, 29
Public repositories, 30, 33
Python, 160–162, 165
  implementations, 161

Q

QoS factors, 125
Q-phase component, 81
Quadrature
  amplitude modulation (QAM), 71, 72, 81–84
  phase-shift keying (QPSK), 72, 81, 83, 84

R

Racism, 62
Radicalization, 62
Random forests, 64
Ratings of alternatives (RoAs), 97, 99
Rational upper bound (RUB), 105, 106, 118

Raw
  data, 201
  materials, 98
RC5 algorithms, 84
Real-life
  domains, 125
  scenario, 19, 153
Real-world
  application, 6, 25
  problems, 24, 25
  scenarios, 23, 172
Recommendation system, 209–212, 215, 222
Reconfigurable
  computing system, 69, 70
  partition (RP), 76, 84, 85
Rectified linear unit (ReLU), 140, 143, 162, 174, 185, 189
Recurrent neural network (RNN), 62, 65–67, 138, 155, 179, 180
Reference summaries, 145, 146, 148, 149
Relative value, 216
Remote sensing, 23, 26, 28
Resource management, 124
Retrospective image registration evaluation (RIRE), 29
Risk
  management, 226, 227
  prediction model, 197
Robotic navigation, 2, 14, 20, 21
Robust synchronization methods, 42
Robustness, 24, 35, 55
Runtime reconfigurable encryption algorithms, 84

S

Saturation point, 238
Scarcity, 5
Search space schemes, 26
Secure communication, 41, 42, 55, 56, 57
Self-adapted EAs, 24
Self-adaptive optimization method, 30
Self-organization, 6, 18, 19
Self-organized systems, 20
Self-organizing maps (SOM), 169
Self-tuned
  approach, 25, 28
  EAs, 24, 36

Index

Self-tuning
  approach, 23, 25, 30, 36
    Cocianu and Stan's method, 32
    De Falco et al.'s method, 30
    Li et al.'s method, 31
    Santamaría et al.'s method, 30
  IR methods, 25
  RIR methods, 34
Semi-circular fuzzy variable (SCFV), 105–118
Semi-static parameters, 226
Sender-recipient relationship, 241
Server energy efficiency, 126
Sexism, 62
Sigmoid, 160, 162, 185
  function, 159–161, 163
Signal analysis and machine perception laboratory (SAMPL), 25, 30, 31, 33
SignNet, 173
Similarity measure (SM), 89, 90, 92, 93, 95, 97, 101
Skip-gram
  approach, 191
  model, 189, 190
Slave system, 52
Social
  agents, 1, 2, 9, 19
  animals, 1, 5
  insect allegory, 1
  media, 59, 60, 62, 67
  networks, 62, 240
Socio-environmental factors, 226, 235
Socio-political behavioral pattern, 2
Soft computing (SC), 23, 24, 107
Softmax, 139, 140, 143, 163, 174, 185, 186, 189
  function, 139, 140, 163, 186
  layer, 174
Software as a service (SaaS), 122
Spatiotemporal crime, 65
Stagnation, 130, 131
Stakeholders, 225, 226, 229
Standalone design, 71
Standard deviation values, 34
State-of-the-art (SoTA), 23, 25, 26, 30–32, 36, 179, 181
Stigmergy, 6, 13, 19

Stochastic
  gradient, 168, 180
  optimization technique, 7, 8
  structuring practice, 1
Structural modelling, 204
Stylometric
  data, 185
  features, 179
  method, 192
Subpopulation, 31
Sub-word information, 179
Superintelligence, 6
Supervised learning mechanism, 5
Support vector machine, 179
Surface
  color discontinuity, 17
  normal discontinuity, 17
Swarm, 2
  algorithm, 129
  intelligence (SI), 1–3, 5–8, 14, 17–21, 129
    advantages and limitations, 17
    introduction, 5
  population, 129
Synapses, 154
Synchronization, 1, 41, 42, 53, 55, 56
Systematic learning, 4
System-on-chips (SOC), 71

T

Tabu list, 16
Tanh function, 161
Taste communities, 3
Telecommunication network, 18
Telemedicine, 171
Text summarization, 135–137
Threshold, 17, 26, 33, 190, 242–244
Tractability, 24
Traditional
  ants, 130, 131
  FPGA implementation, 87
  overview, 73
Transceivers functional blocks, 72
Trapezoidal PFS (TrPFS), 92, 93
Traveling salesperson problem (TSP), 1, 2, 12–15, 20, 21, 130
Trial-and-error method, 123

Triangular PFSs (TPFS), 91, 95, 101
Trigrams, 140
Tuning methods, 33, 34
Two-dimensional array, 16
TwoFish, 72, 84, 85

U

Ultraviolet (UV), 174
Uncertain number, 107
Unigram, 61, 140
   precision, 140
Unique binary data, 81
Universal design, 25
User-based tuning approach, 28
   García-Torres et al.'s methods, 28
   Ma et al.'s method, 28
   Panda et al.'s method, 29
   Yang et al.'s method, 29

V

Variable neighborhood search (VNS), 30
Variance (V), 105, 106, 114, 115, 117
Vector, 62, 136, 138, 139, 142, 148, 179, 180, 182, 185–187, 190, 199, 242, 243
   space, 179, 180
Vectorization, 63, 142
Vehicle routing, 13, 14
Velocity, 8, 128, 129, 131, 197
Virtual
   environment, 11
   input-output (VIO), 84, 85
   machine, 121–126, 132
      optimization, 121
      placement (VMP), 122, 123
      scheduling, 122
Virtualization, 121, 124, 126, 132
Visual assessment, 15

Voice assistance, 3, 4
   Alexa, 4
   Corona, 4
   Google assistant, 4
   Siri, 4
Vulnerability, 63, 226–229, 231, 233–235

W

Waiting time, 237, 238, 245, 247
WaitNet, 173
Wallace tree multiplier, 77
Weights
   of criteria (WoC), 97, 99
   vector, 97
Weighted normalized fuzzy decision matrix (WNFDM), 98, 101
Wireless communication, 71
Word
   embeddings, 138, 148, 180, 193
   vector
      generation method, 186
      map, 187
Word2Vec, 138, 139, 180, 183, 186
Words and Skip-gram model, 138
World
   War II, 141, 143, 146
   Wide Web, 61, 67, 208

X

Xilinx Vivado, 76, 78, 80
   software, 70, 71

Z

ZedBoard, 71, 72, 79, 81, 84–86
Zero-padding layer, 142
Zynq evaluation and development kit, 86